Michelangelo's Finger

Michelangelo's
Finger

An Exploration of Everyday Transcendence

Raymond Tallis

ATLANTIC BOOKS
LONDON

First published in hardback in Great Britain in 2010 by Atlantic Books, an imprint of Grove Atlantic Ltd.

Copyright © Raymond Tallis 2010

1 2 3 4 5 6 7 8 9

A CIP catalogue record for this book is available from the British Library.

ISBN: 978 184887 119 9

Printed in Great Britain by the MPG Books Group

Atlantic Books
An imprint of Grove Atlantic Ltd
Ormond House
26–27 Boswell Street
London
WC1N 3JZ

www.atlantic-books.co.uk

Dedicated to Edna Turnberg (who always sees the point) with love and admiration and with gratitude for being such a wonderful friend

Michelangelo's Finger

Contents

Acknowledgements

It is a huge privilege and pleasure to be working with the Atlantic team for a third time. I am enormously grateful to Toby Mundy, Chairman and Publisher of Atlantic Books, for his enthusiasm for this book, for understanding what it was really about, and for a conversation which not only gave *Michelangelo's Finger* its title but also pointed it in the right direction. My thanks also to Sarah Norman for her excellent editorial work.

Foreword

It is easy to underestimate the influence of small things in determining what manner of creatures we humans are. Over time, the repeated and multiple effects of a slight difference can make a big difference. The independent movement of the index finger is one such small and easily overlooked thing, and it has made a big difference. We sometimes need thinkers of genius to make us see this. Michelangelo was such a thinker, although he usually thought with a paintbrush and chisel rather than a pen. *The Creation of Adam*, one of his frescoes on the ceiling of the Sistine Chapel in Rome, is one of the most familiar images in Western art, depicting, if you believe the story, the most important event in the history of the universe. Yes, God had been pretty busy up to that moment. In just five days he had instructed the void to shape up; had commanded light to come into being and stand in tidy rows of days and nights; had divided the water from the land and heaven from the earth; had summoned grass and beasts and the sun and the moon and the stars into being and instructed them to take up their stations and carry out their proper functions; and had checked, and found, at intervals, that the results were excellent – or good enough anyway for Him to rank them as good. But now we come to the climactic moment, recorded in Genesis 1: 27: the culminating act of creation, at the end of the sixth day, when God played his master-stroke and created man in His own image. And it is this that Michelangelo represents on the ceiling of the Sistine Chapel.

At the centre of the picture are two fingers separated by a small gap: the index finger of God's right hand; and the index finger of Adam's left hand. This extraordinary fresco is open to many interpretations. We might see it as God's outstretched right hand transmitting the spark of life from Himself to Adam, whose left hand is extended in a pose that mirrors that of the Creator. Through His extended forefinger God infuses his spirit into Adam, and hence into humanity. This image is somewhat at odds with the more detailed account of the creation of Adam in Genesis 2: 7, in which God 'formed man of the dust of the ground, and breathed into his nostrils the breath of life; and man became a living soul'. Alternatively, therefore, given that this is an act of creation, we might conclude that it represents the moment *after* the separation of God and man, of the Creator and the pinnacle of his creation. The index fingers are the final point at which separation takes place – a profoundly tragic moment, given what was to happen in Genesis 3, when God is betrayed by the one creature whom He created in his own image.

This leaves the fact that the index finger is centre stage still to be explained. This was no mere eccentricity. The best-known and most venerable of Catholic hymns, sung at supremely important occasions such as the election of Popes, already over 700 years old when Michelangelo deployed his brush in the Sistine Chapel, is 'Veni Creator Spiritus' – 'Come Creator Spirit'. In this hymn, the Holy Spirit is called 'the Finger of the Hand Divine' and the digit in question is the index finger. There appears to be a deep symbolic connection between this finger and the special nature of human beings, who are understood in the Judaeo-Christian tradition as being created uniquely in the image of God. Why?

I ask this question in full awareness that, if the hand has lifted man above other living creatures, the credit would seem to lie with the thumb. Indeed, in *The Hand: A Philosophical Inquiry into Human Being* I argued precisely that.[1] I examined the unique versatility of the human hand, originating from the possession of a fully opposable thumb, that transformed this organ into a tool; and the 'toolness' of the hand altered our relations to our own bodies and ultimately to our environment. Hominids developed an *instrumental* relationship to their bodies and for this, and a variety of other reasons, they were no longer just organisms *living* their lives but self-conscious, embodied subjects actively *leading* them. The crucial role of the hand, and indeed the importance of the opposable thumb, had been postulated by many other thinkers – philosophers such as Anaxagoras, Aristotle, Kant and even Heidegger, and theoretical biologists such as Erasmus, Darwin, and F. Wood Jones, who famously said that 'Man's place in nature is largely writ upon the hand.' In *The Hand*, I took this story a bit further by teasing out how the possession of a uniquely versatile manual organ enabled hominids to become the self-conscious agents that we are; how it provided a biological means by which we loosened the grip of biology and came to live in a world that was increasingly human.

The present examination of the index finger – more precisely, one immensely important function of that finger – takes this story further. Of course the index finger works with the thumb and with other digits to make possible the hand's dazzling virtuosity. At the most obvious level, all of those so-called 'pinch grips' which enable us to manipulate the material world with such exquisite precision involve the index finger. Furthermore, many of the instruments that enhance our precision grip on the material world – needles and screws and so on – would have had no rationale without its contribution. And the index finger also throws in its lot with its fellow digits to give strength to those 'power grips' that are so important to our manipulation of the material world, as when we squeeze objects (including the throats of our enemies) or hang on for dear life. But in addition to this cooperative activity of the index finger, there is stand-alone activity, made possible by its unique capacity for independent movement.

Look at your index finger now. Waggle it about and note how easily it does its own thing, how much more fluid and liberated it is compared to its fellow digits. The others can do similar things, but with more effort and less grace, as if they were merely imitating the index finger without fully knowing what they are up to. Not for nothing is the index called 'the forefinger'. At any rate, it is the one we most naturally deploy when we want to winkle things out of small spaces within and without the body. (The reader may wish to be spared examples of the former and that wish will be respected.) But there is an even more important function, one which it does not share with other primates.

Quite likely when you were invited to look at your index finger just now, you got it to do something that is connected closely with its name: you got it to indicate; that is to say, to *point*. Pointing is a beautiful gesture. But it is much more than this. And it is through pointing, I will argue, that the index finger has contributed so much to hominid development and to the creation of a human world outside of the natural world that encloses all other animals. I like to think that the slightly awkward encounter between God and man through their index fingers depicted by Michelangelo, and indeed the theological idea behind it, was influenced by an intuition of the central role of the index finger in making us so different. The mutual pointing of the index fingers of God and Man was placed at the centre of a supernatural image of man's extra-natural nature.

My own viewpoint, incidentally, and one that I have elaborated in many books, shuns both supernatural and naturalistic accounts of human beings. I believe in what I have called 'Darwinism without Darwinitis'. That is, I do not doubt that we are descended from hominids and that our hominid ancestors came into being by the same processes that gave rise to centipedes, frogs and monkeys. But I have equally no doubt that, since the hominids forked off from the other great apes, they have taken different paths, and their journey has been powered by different processes, from all other living creatures. We did not fall from the sky, or come into our distinctive being by means of supernatural intervention: it is biology that gave us our passport out of nature to a place from which we can manipulate nature to our advantage and to ends that nature did not envisage (not that nature does any envisaging). This is a process I have described in great, indeed pitiless, length in a trilogy of books that the reader may wish to consult but will be spared here, though I will visit the relevant aspects of it in Chapter 2.[2]

The role of the index finger in this remarkable story is by no means insignificant, even though the item in question is not terribly impressive. As is so often the case, it is not the kit which nature has bestowed upon us but what we do with the kit that makes the difference, or, rather, that makes a large difference out of a small one. This is most obviously true of the fully opposable thumb. The human pincer grip, with thumb-pad to finger-pad contact, is very rarely used in chimps who cannot achieve full opposition.[3] This has not only increased the versatility of the hand – even adult chimps, in contrast with infants, lack a means of selecting customized grips for small objects – but has also conferred upon us a more explicit sense of our hands as tools. This lies at the origin of our awareness of ourselves as (self-conscious) agents. It is against this background that we are able to transform the biological givens of our organic bodies into quite other things that serve quite other purposes, something that I have explored at length in *The Kingdom of Infinite Space*.[4]

Making a big difference out of a small one also applies to the index finger and those small developments that have given the forefinger freedom to operate in a way that is not narrowly, or even broadly, prescribed by nature. It is against the background of a hand (and indeed a body) that has become the primordial tool, or tool-kit, of a self-conscious agent that the index finger is utilized as one of the richest means of communication that humans possess outside language. Indeed, pointing is possibly one of the bridges to language (though this claim will be hedged about with qualifications). At any rate, the unnatural nature of pointing and what it tells us about ourselves are the theme of this book.

Just how extraordinary – and indeed extra-natural – pointing is will become evident when, in Chapter 1, I examine what successful pointing requires of us. This will be elaborated in the Chapter 2, where I look a bit more closely at the mode of consciousness necessary to want to point something out to someone else, or to understand what is meant when someone else points something out to us. This mode of consciousness is not achieved in animals, which makes the claim that some animals really do point of considerable interest. In Chapter 3 I examine, and reject, this claim, which reflects a tendency to attribute mental capacities to beasts, an anthropomorphism ubiquitous in research into animal behaviour, especially into the behaviour of primates. That some human beings do not point is itself a pointer to a profound problem: the failure to point is an early sign of autism, a condition that affects language, behaviour and every aspect of socialization and which appears to be founded in an impaired sense of self and of others as selves. Autism is discussed in Chapter 4.

Pointing is often seen as a bridge between the pre-linguistic and linguistic states of humanity. We should beware of this assumption, not only because, as I have already argued, the claim of pointing to be 'natural' in the required sense is ill-founded, but also because the relationship between words and the contents of the world is not one that can be conveyed by pointing. The limitations of teaching names by pointing at objects and of so-called 'ostensive definition' cast a sharp light on the extraordinary nature of language. Chapter 5, which deals with this, includes forays into the philosophy of language which some may find quite hard going but, I hope, worth the effort. Chapter 6 offers the reader something of a break, with a series of observations about some of the things the index finger gets up to that are directly and indirectly connected with pointing, and recalls some of the uses to which the extended index finger may be put that go beyond a declarative pointing that informs another person of something they may not, but should, know. By pointing to others, we also increase their self-consciousness, and this is not always neutral or benign. The reasons why it is rude to point, and how pointing may be a way of asserting power, touch on something close to the heart of what it is to be a human being. Chapter 7 deals with assisted pointing; with the prostheses and prosthetic extensions of the forefinger that human beings have manufactured to perform indication more effectively than with the forefinger or even in the complete absence of a conscious indicator. The metamorphoses of 'the pointer' are legion and the transformation of an act of pointing into a standing artefactual pointer is deeply thought-provoking, reminding us of how we turn an infinity of strangers into a cognitive community.

The final chapter delivers the philosophical message of *Michelangelo's Finger*. Pointing is a means of indicating a transcendent world – general, hidden and shared – and takes us decisively out of our solitary, transient bodies, subject to the laws of nature. The examination of this aspect of pointing begins with the transcendence inherent in everyday human consciousness, takes in my own boyhood and that of Sir Walter Raleigh, and ends with the idea of God – a God whom we see pointing back to us and thus bringing our distinctive nature into being. In short, I end, where I began, with Michelangelo's vision of humanity, though I give it an interpretation of which the Catholic Church would most certainly not approve.

I first became interested in pointing in the early 1970s and in 1973 I wrote a first draft of this book, under the title *Studies in Pointish*. I laboured over it, in the spare time I had from my 104-hour week as a junior doctor, for the best part of a year. I then typed it out and, as this was before the era of word processors, I typed it out again.

And again. When, finally, I had produced a script that was not so bespattered by Tippex as to look as if it had been composed under a flock of seagulls, I sent it off. A succession of extremely patient editors and publishers found it unsatisfactory. And so, as I in the end admitted to myself, did I. Pointing excited me, but I did not know why. I had to write the trilogy referred to earlier to see the bigger picture into which pointing fitted and, indeed, pointed to – an illustration of how we may learn as much from writing books as from reading them. Finally, a conversation with Toby Mundy, of Atlantic Books, in which he suggested the title *Michelangelo's Finger*, made me think a little harder and deeper and really see what my excitement over this exquisitely beautiful – and terrifying – gesture was all about.

As much as anything else, *Michelangelo's Finger* is a contribution to the unfinishable project of waking to, and out of, the enabling constraints of everyday meaning, of linking our ordinary moments with their extraordinary origins. The fact that I have focused on something that seems trivial and far less worthy of examination than language, or human consciousness, or the working of the brain, is no accident. Pointing (to pick up on one of an endless number of puns waiting in the wings) points to something that goes very deep in us; and it is subject, as so many things in human life are, to a multitude of transformations. I hope this inquiry into the nature and significance of, and the role played by, this small gesture will make that point, and help us to see ourselves more clearly.

The fact that humans are the only creatures who routinely point things out to one another links, ultimately, with what is distinctive about our nature, and that, irrespective of whether we are made in the image of God (whatever God is), we are not such as is mirrored in nature. Indeed, we are a mirror in which nature sees herself. Because we transcend our natural condition, we are aware of our own nature and of nature herself in the way that no other part of nature is aware. Pointing both presupposes and develops that transcendence. I was tempted to call this book *The Godfinger*. I believe, however, that, in placing the forefinger of God and the ultimate forefather of man at the centre of the act in which humanity was created, Michelangelo captured a great truth – a greater truth than he perhaps realized.[5]

chapter one

How to Point: A Primer for Martians

There can be few more dispiriting experiences than being the recipient of detailed but entirely superfluous explanation. Of all the things readers of this book may feel they need, instruction in how to point might seem to be the least pressing. But it is a necessary step towards understanding what pointing, ultimately, points to.

At first sight, nothing could be more straightforward, natural and unpuzzling, than pointing. It seems, of all our gestures, to justify what St Augustine said when he famously described bodily movements as 'the natural language...of all nations'.[1] Pointing appears to be the least conventionalized of the signs human beings use to communicate with one another, and consequently to require the minimum of decoding. It certainly appeared so to the Sophist Cratylus. According to Plato, he argued that, if we communicated solely by pointing, misunderstanding would be avoided.[2] Indeed, 'Pointish' seems so transparent a language, or proto-language, that it has frequently been seen as the key to the miracle by which the speechless infant becomes the toddler who speaks. And when some-one wants to know what you are talking about, you can always, as a last resort, point to it.

However, all is not what it seems. Pointing is not at all straight-forward. What is more, it does not deliver what would be needed if it were the sole bridge from babbling babyhood to talking toddler-hood, or an all-purpose means of clarifying what is meant when language fails us. Nevertheless, it is central to developing the mode of consciousness – explicit, shared, collective – that is infinitely elaborated in (uniquely human) language. It is worth dwelling on this a bit.

First of all, as just noted, pointing does not always deliver what is required. The index finger by itself is not sufficient to make clear what it is that is being pointed to, at, or out. There is the well-known and probably apocryphal story of the anthropologist who wants to learn the language of a newly discovered tribe. Accompanied by his native informant, he points to a series of pictures he has brought with him – of objects such as a dog, a house, a tree and so on. To his astonishment, each picture elicits what sounds like the same word. He gets very excited. What kind of world-view must these people have, if they use the same expression for things as disparate as dogs, houses and trees? Has he stumbled upon a new mode of human consciousness? In the middle of the night, the penny drops. The word elicited by pointing to each of the pictures is the word for 'picture'. And non-apocryphally, David Wilkins reports that he had to be retrained how to point 'properly' when he lived among the Arrernte people, in particular to appreciate the key role of pointing with the extended lower lip to supplement index-finger pointing.[3]

Secondly, if pointing is a 'natural' mode of signification, it must be a rather special one, because humans are the only living creatures who point. This is why nothing is straightforward about pointing: it partakes of the complexity of human consciousness and, indeed, I argue that it has played an important part in its elaboration. Before we examine these complexities, we need to get our object more sharply in focus.

What is pointing? It is sometimes a wise, and always a safe, move to consult a dictionary to find out what you are talking about. When the lexicon in question is *The Oxford English Dictionary*, the reward is generally beyond expectation, and a search for definitions rapidly turns into an archaeological dig into the accumulated past of human thought. (Of this, more presently.)

A Lexical Interlude

'Point' has so many primary and derivative meanings and uses that we are in danger of losing ourselves in a labyrinth. The *Oxford English Dictionary* lists sixty-five meanings. That to which we ultimately point, what the grammarians would call its cognate object or internal accusative, is a 'point' and this itself proliferates like the hyphae of fungi. A 'point' may be (to pillage the *OED*) a prick (absence) or dot (positive). It may mark a pause to articulate the sense of something, giving notations or time junctures, hence decimal points and the points seen in mediaeval musical notation, indicating a note, that is to say a separately produced sound, though it would be several hundred years before point would acquire another meaning in opposition to counterpoint. It may, of course, mark the end of a separate piece of sense or communication – a period. It is notionally a minute particle of anything, the smallest unit of measurement; the smallest or a very small portion of time; and ditto of space. Points may be units of counting, as in scoring games, competitions or examinations. They may mark accumulated credit in credit cards; may be the measure of the size of typeface (this book has been typeset in 12 point). Most mysteriously and paradoxically, they may indicate position without magnitude. The point may be: the precise matter being discussed ('Your point being?' – the perfect put-down); that at which one aims, or for which one strives or contends – aim, object, end; a conclusion, culmination, period; or a sharp end to anything. And expressing something with delicacy, we put a fine point 'on' something. An indication, a hint, a suggestion may be all that is necessary to re-locate a certain fieldsman in cricket from point to silly point or to criticize the extremities of a horse.

The verb 'to point' has no less than sixteen stem meanings, several of them with numerous branches. The one most pertinent to our inquiries is the ninth: '9a. *Intransitive* To indicate position or direction by extending the finger; to direct attention *to* or *at* something in this way.' That is what pointing does. How does it do it? What would one tell a Martian about the basic rules of Pointish? To indicate something in the canonical form of pointing, you need to extend your arm and index finger in the direction of that which is pointed at. Precision pointing requires the index finger to be offset from the others and it is relevant therefore that in humans the index finger naturally protrudes above the other fingers, whereas in the non-pointing chimpanzee it does not. Aside from the ability to isolate the finger, pointing is more of an achievement than it sounds. My years looking after patients with neurological problems have been a constant reminder of just how much of an achievement accurate pointing is.[4]

How to Point: Physiology and Biomechanics

Point at an object. Observe yourself extending your arm and index finger in the direction of the target. Most likely, you will have separated the signal of the index finger from the noise of the rest of your hand, to ensure precision, by curling the other fingers under the palm of your hand. Your thumb will assist in this by pressing on the middle finger, as if holding it back. The arm has to be held steady, so that the long axis of the index finger and the imaginary line connecting you and the object are congruent. Stretching out your arm involves muscles around the elbow and shoulder joint acting in coordination, so that there is a smooth unfolding. Maintaining the position requires the careful calibration of the force exerted round the elbow, so that the extended position is maintained. The position of the elbow as a whole has to be sustained by the operation of a galaxy of muscles in the upper arm, in and around the shoulder joint, and even in the trunk to maintain the stability of the shoulder joint, and to keep the shoulder itself in place. The necessary stability to maintain fixation on the target, in short, requires the exquisite control of numerous muscle groups acting in concert. This control – from the fractionated finger movement that separates the index from its fellow digits, to upholding the outstretched arm – is made poignantly evident in its absence, for example, in patients who have had strokes, who cannot separate their curled-up fingers, move their arm into the right place, or hold it steadily once there.

The achievement inherent in judging the appropriate position of the outstretched hand is easy to overlook. The object is over there and the arm is simply invited to, as it were, engage in a 'virtual reach'. But assuming the right posture in the right place is not as easy as it appears. First of all, each of the joints – the shoulder, the elbow and the wrist, has several degrees of freedom: there is a range of possible positions it can adopt in several planes. The choice of positions has to correspond to the relation between three things: myself, my arm and the object to be pointed out. Pointing, in other words, has to be enacted within a frame of reference which defines the coordinates of myself and of the object, which will then in turn define the location and characteristics of the line that will link the one with the other. I locate myself at a kind of 0,0,0 point within an egocentric space that encompasses both me and the object. The challenge, therefore, is then to translate the relation I see, and feel, between myself and the object, into a line that links them; and use my own arm to flesh out (literally) that line, so that it is visible to the person for whose benefit I am pointing. In order to do that, I have to translate the difference between the present position of my arm and the position necessary to point at the object into patterns of muscle activity.

There is much more to it than I have described but I imagine I do not need to say more to persuade you of the scale of this achievement. It is possible to point of course with the whole arm – and people who have had amputations do that – but fine-tuning requires the separation and unfolding of the index finger. This is an example of so-called *fractionated finger movements*. These are controlled by a particular pathway in the nervous system, the cortico-spinal tract which is a hotline between the cerebral cortex and the nerves in the spinal cord controlling voluntary muscles. Other primates have fractionated finger movements controlled by the cortico-spinal tract, which enable them, for example, to winkle nuts out of tight spots; but these movements are far less well developed than in humans. Just as chimpanzees do not have a fully opposable thumb, so their ability to separate their index finger from its fellow digits is less well developed.

Reaching, holding the extended arm steady, against gravity and other countervailing forces, and even fractionated finger movements, are not, of course, unique to pointing. And the use of the index finger in the way described is not the only mode of pointing, although it is the canonical referential gesture that makes clear what is present in other, less versatile, modes of bodily pointing, using the thumb, the arm as a whole, the elbow, the shoulder, the head, the torso, the eyes and even the foot. Index finger pointing makes most explicit the essence of pointing.

How to Point: The Rules of the Game

What *are* unique are the rules of the pointing game? It is quite a business and there are four components. There is the *producer* (the person doing the pointing); the *pointer* used by the producer (usually the outstretched hand and index finger); the *pointee* (that which is pointed out); and, finally, the *consumer* (the person for whose benefit the pointing is carried out). The producer uses a part of his or her own body to establish an axis that joins the producer with the item being pointed out – with the pointee. The consumer is invited to follow the virtual line with her visual attention until it reaches the pointee.

The nature of the pointee may vary enormously. It may be a material object; or the rough location of an object (as when we point 'somewhere over there' or at the clouds in the sky or 'over them hills yonder'); or the direction of an object whose precise location is or is not known; or simply a direction, as when one is pointing to which way someone went, or where something was located a little while back (as when, for example, I point to an empty chair), or indicating in which direction a town is, or which way is north. In the most straightforward case, the pointee is a determinate, clearly located, clearly defined material object – a cup, a cat, the kind of thing that the philosopher J. L. Austin, with tongue slightly in cheek, would have called 'medium-sized dry goods'.

The pointee gets pointed out in virtue of lying on a line projected from the tip of the producer's finger. The outstretched finger lays down the first few inches of the line by embodying it. The arm of course is not merely the transporter, the supporter, the plinth, but often forms part of the line itself. With its help, Pointish is whole-arm-shouted rather than finger-whispered. This is important when producer and consumer are separated by distances that shrink the phenomenal appearance of their bodies – as when both are in the Great Outdoors.

So much for the producer, the pointee and the pointer. What of the recipient or consumer, without whom pointing would be point-less? To benefit from the producer's action, the consumer has to understand the intention behind the action, to understand why the producer is assuming that particular bodily posture. And then she has to work out what it is that is being pointed to. Identifying the act as one of pointing should not be too difficult, if only because there are relatively few reasons for holding up one arm and making it and the index finger stick out in what we might call the 'indicative' mode or mood. Pointing postures are assumed only for the sake of point-ing; they rarely happen accidentally or as part of other actions. Even so, it is not self-evident that the hand-and-finger is indeed pointing rather than merely reaching out or signalling in some other way. While the very name of the index finger broadcasts its distinctive role in indicating things, it does not have its name written on it. Grasping (a word whose over-determination scarcely needs underlining) what the producer is pointing at, out or to, however, is even more demanding.

What is so difficult about it? Quite simply this: in order to locate the pointee, the consumer has to assume the position of the producer. Clearly we do this mentally, not bodily, though when pointing fails and we cannot see what the other person is pointing out, we might come close to her and look along her arm as if along a telescope. (A certain amount of irritation is often evident under such circumstances: the producer with the dimness of the consumer and the consumer with the vagueness, or unhelpfulness, of the producer.) Normally, though, the consumer has to cast herself in her imagination out of her own body and mentally look along the line drawn in space by the arm and index finger extending from the producer's body. The consumer, that is to say, has to put herself in the producer's place. This is a rather remarkable thing to do: it amounts to an, admittedly minor and temporary, but nonetheless real, abdication from the sense that one is at the centre of those things that are lit up in one's sensory field; that one is the centre of the experienced universe.

This is an extraordinary achievement, and it says a lot about what we humans are. In the next chapter, I shall argue that this voluntary displacement of the human subject from the material centre of his world is a first step in the growth of an important intuition: that one is part of something greater than one's self and greater than the parish uncovered by one's sense experience; that one is part of an explicit community of subjects; and, ultimately, that one is an atom or unit of a society. Acquiring the ability to point or to understand what someone else is pointing at is a step in the *collectivization* of one's individual consciousness through the joining of attention. Pointing, in short, points towards a distinctively human form of awareness; a uniquely human breach in the solitude of sentient creatures.

It is not always, or even usually, necessary to put one's self *precisely* in the other's person's place in order to identify what it is he is pointing at. Sometimes, as when pointing is supplemented with words – 'Look at that cat!' – it is necessary only to look in the right quarter of space.[5] This composite mode of pointing – fingering plus words – is more advanced, a development occurring later in an individual human life and later in the history of the human race. At any rate, interaction between carnal (digital) and semantic (verbal) pointing, between pointing with the flesh and pointing with air or ink, is surprising, indeed strange. Just how strange this composite pointing is will become clear when, in Chapter 5, we discuss the intersection between semantic (or linguistic) and material (or physical) space in the use of so-called 'ostension' in teaching the meaning of words. For the present we shall accept that the producer may sometimes use words to assist the consumer, so that the latter does not always have to stand imaginatively at the root of the producer's pointer.

The background to what the *producer* does is somewhat complex. She has to be able consciously to use her body as a signal. This implies a special relationship to said body, one that (as we shall discuss in the next chapter) is not found in animals. In addition, she has to have the capacity to be aware of another's (different) viewpoint. This is a necessary condition of her being aware that she is cognitively advantaged compared with the other person, at least with respect to knowledge of the object being pointed at. In addition, she has to understand that the other's comparative disadvantage can be set right. We shall say more about the use of one's own body as a signalling device in the next chapter. For the present I want to focus on another aspect of pointing.

Pointing and the Minds of Others

The point of pointing something out to another is to amend a perceived deficit in their knowledge, or experience, or awareness. There are people who weary us by pointing to things we can already see. And we may also point out to each other things we both know we can both see in order to underline our togetherness. The usual, and fundamental, occasion for pointing, however, is to correct a lack: to draw attention to something important or at least interesting the other has not noticed or cannot see. This is the meaning of the *meant meaning* that is pointing. For pointing does not have meaning by accident; nor is the meaning it has usually accidental. While it is probably correct to believe that the meaning the consumer takes from 'signals' emitted by animals are not meant – they are simply overspills of presence – this is not conceivable in the case of pointing. That which is meant – the pointee that is pointed to – is *meant to be meant*. Producers use their arms/fingers in a deliberate, discretionary manner.

The connection between pointing and the awareness of others' cognitive deficits (their as yet unmet needs for information) is poignantly underlined in the case of humans who do not point, namely individuals with autism. We shall discuss autists in Chapter 4, but it is worth noting that a plausible framework for making sense of this harrowing condition is that autistic individuals are 'mind-blind'.

They do not have the sense that others have a different viewpoint on the world, one which is partly but not completely shared, even when those others are in close physical vicinity. More precisely, they have a very limited sense that others have a viewpoint at all. And they lack this sense of the other's self because they themselves do not have a fully integrated sense of self.

Which brings us to the following definition in the 'point' entry in the *Oxford English Dictionary*: 'Pointing b. *fig* To direct the mind or thought in a certain direction: with *at* or *to*; to indicate, suggest, hint *at*, allude *to*.' This emphasizes the extent to which the producer is concerned with the mind of the consumer. Pointing *at* is pointing *out*. The modulation of 'at' to 'out' requires of the producer that she has a forked consciousness: one tine of the fork skewers the object and the other tine is directed towards the consciousness or attention of the consumer for whose benefit the pointing is being carried out. This forking is echoed in what happens to the consumer: his attention is drawn, by the attention he pays to the producer, to the object.

I think it is worth while to stare a little longer at what is going on here. You are trying to grab my attention and re-direct it towards a particular object. You do this typically by inviting my gaze to point in the direction of the pointee. So first you look at me, to ensure that I am seeing your arm and that, seeing what it is pointing at, and making allowances for the different starting points of our gazes, I re-direct my gaze accordingly. By this means, you get me to attend to that which, courtesy of your pointing arm, you show what you are attending to. Pointing makes the axis of your gaze visible by turning the notional line connecting your eye and the object into a real one by realizing the first part of the line in finger-flesh bolted on to arm-flesh. Your fleshly embodiment of the axis of your gaze looking at the object makes that gaze visible.[6] By means of getting me to share your visual attention you draw attention to the object.

I have used the word 'attention'. Its etymology is of particular interest. The word comes from the Latin *attendere*, which literally means 'stretching towards' from *ad-* 'to' + *tendere* 'stretch'. To attend is to direct the tendrils of one's mind to and around an object. This active reaching of consciousness is made evident in your explicit, visible attending and in the attention that you solicit from me, the consumer. I ad-tend in response to your invitation, insistence or command. I may not at once see what it is that I am supposed to attend to: successfully homing in on the pointee is usually the result of deliberate scrutiny, a search prompted and guided by the pointer, supplemented by more or less irritated explanations.

Less Basic Pointish

Let me end this opening chapter, by very briefly considering a couple of more cardinal senses of 'to point'.

First, the dictionary notes the verb may be used transitively: 'To direct (a finger, a weapon etc.) *at*, to level or aim (a gun) *at*; to direct (a person, his attention, or his course) *to*; to turn (the eyes or the mind) *to* or *upon*.' This warrants some unpacking. We have already covered all that needs to be said about the first component, directing attention. The third, turning mental attention, encompasses both the producer and the consumer. The producer actively attends without external guidance, encouragement or invitation; while the consumer actively attends in response to guidance, encouragement or invitation. The second component, directing an object, in part identifies one element of ordinary pointing: directing a finger *at*. But it goes beyond that. Importantly, it highlights the connection with the use of artificial pointers. In this case, the device that is used to point has more in its sights than merely identifying its object: it threatens to destroy it.

Secondly, 'to point' may be used intransitively: 'Of a line or a material object; to lie or be situated with its point or length directed *to* or *towards* something; to have a specified direction; also, of a house etc., to look or face'. This sense of the verb relies upon our projecting into inanimate entities some part of our own capacity to indicate things. It is an example of how we infuse into – or impute to – material objects actions we ourselves do, or intentions we have. Such transferred animation is the basis of those material pointers (such as signposts) that are capable of stand-alone pointing, acting as our proxies in our absence. (See Chapter 7.) It is part of a wider tendency, central to our folk physics, to attribute agency to material objects, although in a world disenchanted by the rise of science, we call this causation. The 'animist projection' is, anyway, incomplete: we don't really believe that inanimate objects can point. When we speak of a line 'pointing' north, or across the page, we are implicitly imagining our own gaze following a projection beyond the end of the line as if it were a pointer. It hardly needs saying that the pointing lies within the conscious subject not within the line, which simply exists, and does not deploy its own long axis to make a direction explicit. (And, indeed, if it did, it would have to point north and south simultaneously.)

The projection of intention from a conscious subject is especially obvious when we speak of a blunt object – say a house – pointing in a specified direction. In order to get a house to point, we have to infuse it with a proxy consciousness: we say that it 'looks' or 'faces' in that direction. Houses, of course, don't look and they do not have faces, not of the kind that look or are looked out of. The metaphor is deliciously complex. First, it transfers to the house the consciousness of someone who goes into, and looks out of, it. Secondly, it makes directed attention – the gaze of someone looking out of its 'face' – a kind of pointer. The transferred epithet implicit in the notion of a house 'facing' in a particular direction is a particularly engaging example of a ubiquitous tendency to animate the world with our own minds.

Time for one final observation on Pointish. While the simultaneous extension of the arm and index finger towards a target is the usual mode of carnal pointing, it is not the only one, as already noted. There is hardly a part of the body that cannot be used as a pointer. Pointing with one's eyes, supplemented by tilt or twist of the head, is particularly common. When someone looks in a certain direction, we are often prompted to look in that direction ourselves, adjusting for the fact that our zero point is different from theirs, especially if we feel their gaze has been attracted by an object of overwhelming interest. The pointing glance – or pointed glance – is most often deployed when we want to warn someone of the presence of a third party, without making that third party aware of the warning. Suppose I am making some scurrilous remarks about Jim to Jane. Jane thinks that Jim may be within earshot, and is exquisitely embarrassed, but does not know how to warn me. She looks away from me in the direction of Jim. I will be drawn to see Jim because Jane's act of looking is a bit odd: normally when we are talking to people, they reciprocate our gaze. Jane may italicize the point of her pointing gaze by staring hard at me or even opening her eyes more widely than the context would seem to require. The entire head may be used in societies where taciturnity is valued as a marker of masculinity. 'Which way did them cheating liars vamoose?' I ask a cowboy in the street. My interlocutor ejects a jet of tobacco and saliva and nods in the direction of the dusty road out of town.

The discriminative capacity of ocular (eye) or cephalic (the whole head) pointing is grossly inferior to that of brachio-digital pointing. It is also clearly a secondary development, just as the use of other parts of the body, such as the head or the leg, as explicit instruments deployed in a discretionary fashion, is secondary to primordial 'tool-ness' of the hand. This should alert us when we think that we have come across examples of pointing in the animal kingdom, most notably in certain breeds of dogs. But before we come to consider these apparent examples of animal pointing, it is necessary to examine the relationship between pointing and the modes of con-sciousness that are uniquely developed in human beings. For this is one of the key motivators of my inquiry into this deeply mysterious tactile glance, this virtual touch, this pre-linguistic mode of commu-nication that plucks at our attention.

chapter two
What it Takes to be a Pointer

More than a Beast

In case the reader is starting to suspect that she has fallen into the hands of a park-bench crank, obsessed by some trifle that most people justifiably disregard, I must emphasize that I am not alone in finding pointing fascinating. Linguists, semiologists, anthropologists and others have been drawn to this gesture. We have already seen how its apparent simplicity is deceptive. Pointing has three fundamental characteristics: it is dialogic, since it is (usually) for someone else's benefit; it serves to single out an entity which the recipient understands to be the referent; and it defines the direction of the referent as being away from the pointing hand, along an axis defined by it.[1] In the previous chapter, I spelt out the kinds of things that Martians would need to appreciate in order to be able to join in the pointing game. I hope that this approach has turned up enough unexpected, or at least overlooked, aspects of pointing to justify itself, because I now want to dwell a little longer on what we might call the philosophical anthropology of pointing; on what it is about human beings that makes them pointers, or, to turn this on its head, what our pointing behaviour says about us. I believe that pointing is an expression of something distinct and fundamental in our consciousness.

Producers, people who point, we noted, have two interesting characteristics: they utilize their bodies as instruments of communication or, to put it more precisely, they deploy the appearance of parts of their bodies as signals; and, intimately connected with this, they have an awareness of the minds of their fellows and of the information that those minds might lack at a particular time. The former enables them to point and the latter explains why they should take the trouble to do so. Behind pointing, in short, is an explicit awareness of one's own body and an explicit awareness of another's mind. We tend to take both of these things for granted, but we should ask ourselves how they are possible. I think both are based on the way in which we are conscious of ourselves, the manner in which we are (to use a phrase introduced by a truly great twentieth-century French philosopher, Maurice Merleau-Ponty) *embodied subjects* rather than organisms.[2]

Human beings enjoy, and suffer, a self-consciousness that is unique in five respects: it is uniquely sustained; uniquely complex; unique as to its contents; uniquely stitched together internally both within a moment and between moments; and uniquely interwoven with the self-consciousness of others. Those who wish to narrow the gap between man and beast – under the mistaken impression that this is what commitment to Darwinism requires of us – challenge this claim. They note, for example, that some animals – higher primates and elephants – seem to recognize their own images in the mirror. In an endlessly cited experiment reported over thirty years ago, George Gallup observed that when a chimpanzee whose face had been marked with a red spot was placed in front of a mirror, it tried to remove the mark by rubbing its face.[3] This suggested that the beast was aware that the image in the mirror was of itself and that this self was identical with its own body. In short, it appeared to have some kind of self-consciousness. Well, yes; but such episodic bodily self-awareness falls far short of the complex, multiple narratives of the self-awareness of humans; the lifelong sense of a self that is internally connected and forms the basis of a life that is actively and purposefully led rather than one that is merely lived.

In the history of hominids, and in the history of each one of us at the start of our development, the awakening and elaboration of self-consciousness begins with the sense that we *are* our bodies. We begin roughly where chimpanzees begin; the difference is that in the case of chimpanzees this initial intuition is dim and does not develop. What I have called the 'existential intuition', which is full-blown only in humans, the sense each of us has that 'I am this' has, as its first object, or strictly, its complement, the living body. The intuition that 'I am this body' is complicated by the fact that I only incompletely colonize the body that I am. There are many parts of Raymond Tallis's body that I am only intermittently aware of; and other parts that I am aware of not at all, though they are continuous with the parts that Raymond Tallis counts as himself. The parts that I am aware of and count as myself vary, as I am aware of different parts of my body at different times, often depending on what activities I am engaged in. For example, I am aware of different parts of my body when I am pushing a broken-down car than I am when I am sitting thinking. What is more, my relationship to those parts that I am aware of varies, in an infinitely graded fashion, from that of being identical with them, to suffering them, to having them, to using them, to exhibiting them, and to having them as objects of factual knowledge.[4]

No part of the body that I am is completely experientially or cognitively transparent. My immediate experience and my knowledge do not entirely illuminate the material of which I am made. The very entity that I am is only partially scrutable to me. I thus have a feeling, formalized in Descartes' separation of his body from his mind, of being something – a subject, an 'I' – *within* my body that I both am and am not. This extraordinary state of affairs – what Merleau-Ponty called the 'ambiguous' character of the embodied subject – has several consequences that bear directly on our understanding of pointing. There is a strongly developed, explicit sense of objects 'out there', in opposition to, or facing, myself 'in here'. The embodied subject feels itself to be opposed to the material objects around him. We have an explicit sense of ourselves as being *here*, located within the sensory field around us populated by objects, as being an object among objects, being and having a point of view. We 'stick out' in the world disclosed by our sensory experiences as objects in our sensory field as well as being the implicit or explicit subject of that field. This complex intuition opens on to the sense of another individual having a viewpoint – that of an object like ourselves, an embodied subject – that is different from our own.

Hand-Made Humanity

This is what we might call the 'deep background' of pointing, the difference that makes us the only animal that points. How did we get to be so different? It is a complex story and the things that led us to wake up to ourselves – that led to 'I am' and 'We are' and to a world

'That is' – are many and various. They include the upright position, the dominance of vision (for reasons that I will discuss in the final chapter) and, perhaps even more than these, the possession of a full-blown hand that gives us an explicitly instrumental relationship to our own body, a sense that we are agents and that our bodies are the means by which we bring things about. Pointing is a striking expression of this *instrumental* relationship that humans, alone among the beasts, have to their own bodies. The hand that is deployed in classic pointing has several unique features that are relevant to the awakening sense of our bodies as instruments, and ultimately to our sense of the material world as a nexus of causes and effects, enabling us to act on it more effectively.

The most important is one we have already alluded to: the *full opposability* of the thumb to other fingers, in particular the index finger. As a result, the hand has an immense versatility, so that each act of prehension is the result of a choice. The choice is not entirely random: it is constrained by, and gets its meaning from, the material that is to be manipulated and the goal of the manipulation. This 'constrained manipulative indeterminacy' is the seed of the sense of agency, of the explicit sense of using one's own body to achieve explicitly entertained ends, of the notion that the world can be operated on indirectly through manipulating causes of desired effects; and it is the first step towards a life mediated by tools and other intermediaries. There is another consequence of full opposability: there is much more touching of finger by finger in ordinary manipulation. This *'meta-fingering'*, where fingers touch, or finger, each other, raises the self-awareness of the hand, making its status as a tool yet more explicit. Thus, full opposability and all that follows from this, built on to the capacity for fractionated finger movements that is shared with some non-human primates, gives something more than increased dexterity: it lays the ground for the transformation of the relationship to the body to one of explicit instrumentality. This in turn enables the special dexterity of the index finger to be the basis of a mode of communication not seen elsewhere in the animal kingdom.

The emergence of the hand as a tool – 'the tool of tools' according to Aristotle – has had numerous consequences, but two are of particular importance. Firstly, the remainder of the human body also becomes more or less instrumentalized or is available for instrumentalization. Secondly, the intuition of 'toolness' is projected beyond the body. Humans create an environment densely populated with literal tools. The first tools were naturally occurring objects used as artefacts – 'naturefacts'. Subsequently, ever more elaborate artefacts were manufactured with specific functions in mind. Cities are the ultimate expression of tool-making man, being both networks of tools and mighty tool-chests. The spread of instrumentalisation beyond the hand does not, of course, diminish the latter's importance. The hand has remained the ubiquitous, primordial tool through which other tools operate, both the final common pathway for tool use and a tool in its own right. Even those quasi-autonomous tools, machines, require a hand on the tiller or wheel or lever, or a finger on the button.

The hand has not only manipulative functions. It has a crucial cognitive role, being the chief organ of touch. And it is, besides, a major communicator, boasting an almost infinite variety of gestures, one of which is the pointing that is our present concern. Cognitive, gestural and manipulative functions are interwoven: the dexterity of the hand would not be possible without its exquisite multi-modal sensitivity; and, through its gestures, it engages in a special, indirect, 'hands-off' kind of manipulation, acting on the world through acting on others who in turn act upon the world or upon one's self. Pointing is perhaps the most important basic example of this indirect manipulation. In common with all gestures, it exploits the external appearance of the hand, the way it is manifest to others, in order to communicate. Using the hand in this way presupposes a very high order of bodily self-consciousness, at the very least a well-developed sense of oneself as a visible object in another's sensory field. As one who points, I must be aware not only that I am visible to others but also have some idea how I am visible, and how I might utilize what is seen of me. The gesturing hand uses not its power, nor its precision, but its mere appearance, to manipulate the world, via another person.

Just how special the hand is may be illustrated by an experiment that requires no equipment except a human body and a quiet room where one can be undisturbed. Slip off your shirt and let your bared shoulder cool. Touch it with your warm hand. You will find that you are divisible into at least two subjects and two objects. Your hand (subject) is aware of the coolness of your shoulder (object). Your shoulder (subject) is aware of the warmth of your hand (object). There is therefore a double distance within you as an embodied subject. However, this relationship is not symmetrical. The hand has, well, the upper hand: it is manifestly the exploratory agent and the shoulder manifestly the explored surface. Although touch is reciprocated – the toucher in each case is also that which is touched – there is this hierarchy of roles because the hand has come to the shoulder and not vice versa and, what's more, the hand has an established track-record of being an explorer, unlike the shoulder. The differentiation of roles, so that one part of your body is as it were 'superior' to another, maintains the inner distances: the subject–object distance awoken within your body is not cancelled by an equal and opposite object–subject distance. Opposite, yes; equal no. The role of the hand as an exploratory organ as well as a tool creates the sense of the subject within the body, and this in turn confirms and underlines the hand's status as an agent. It does so, however, only once the intuition of agency has already been ignited. That is why its exploratory function in other primates is not sufficient to awaken in them the sense of self and agency. This subject–object, agent–patient relationship within the human body is many layered.

While, as we have seen, other bodily parts may be mobilized to make indicative gestures, the hand is the key organ of indication. It is not only the place where, as we have seen, the instrumental use of the body awakens; it is also especially well placed to signal. The fact that it protrudes from the body and, furthermore, is at a distance – a visible distance – from the centre of consciousness which is located somewhere at or near the head, underlines its general role as a tool and its more specific role as a tool of communication. I see my hand at work – and indeed much of its work is guided by sight – and I see my hand out there as something potentially visible to all. My index finger, at the end of my arm, while it is obviously a part of me, is also differentiated from me.

Pointing is a voluntary, discretionary, act that is susceptible of a multitude of chosen and calibrated variations. This is in sharp contrast with the involuntarily triggered stereotyped bodily postures whose assumption requires no sense of the world picture of any other person and whose variations are random rather than intentional. The use of one's appearance to bring about certain effects, by means of the proto-tool of the hand, is quite different from the reflex or instinctive habit that animals exhibit, of assuming certain appearances. Non-human animals do depend on their bodily appearance for quasi-social purposes, but they do not use this as an explicit *instrument* to bring about certain beliefs in the minds of other animals. For example, an animal that bares its teeth, or raises its hackles to look bigger, or exposes its rear to indicate sexual availability, is not referring to anything outside of itself. There is no differentiation between a subject, a body part used to make a reference, and its referent. The human use of the appearance of the hand, the supreme and original bodily tool, to make something else – an object, a state of affairs – evident to another is entirely different from the instinctual assumption of bodily postures for the sake of their appearance. In using my hand to point out something to someone else, I am taking the division within my body between tool and tool user to a higher degree.

At any rate, the protrusion of the index finger at the end of the arm is an obvious advantage from the consumer's point of view, because it makes the pointer more readily visible. It is equally important from the producer's point of view. The superficial reason is that it permits a great versatility: sticking out from the body, and attached to a hand that is in turn attached to an arm that is in turn fixed to the trunk by a ball-and-socket shoulder joint, the finger can be used to discriminate between a very large number of directions, and hence point to a very wide range of locations. The more profound reason is that, being offset from the self's centre, it is more clearly amenable to being thought of as an instrument which can be used in different ways, including serving the needs of another for information.

Pointing and Human Consciousness

Our unique self-consciousness as embodied subjects underpins our intuition of objects that exist in themselves, independently of our current experience of them, of multiple points of view different from our own, and of others' incomplete knowledge, that together make sense of and justify the act of pointing. Once these intuitions have been awoken, extraordinary developments are possible. The sense that there is more to be known, and that you may be in a position to know it, lies at the root of my intuition of being located in a common world, a world which is not centred on me, and which extends infinitely, or indefinitely, beyond the horizon of my sensory field. This sensory field is abuzz with rumours of a wider field of awareness that belongs to the collective. For modern man, these rumours are mostly word-borne; indeed, we verbally inhabit a world that vastly outsizes the field of bodily sensations whose centre is marked by our bodily being and the notional location of the subject within the body. But this cognitive universe does not begin with words. It is conceivable that it begins with that family of gestures, of communications, whose paradigm example is pointing.

Before I defend this rather bold claim, let me define one or two terms. First of all, 'sentience'. This is a form of consciousness that humans share to varying degrees with non-human animals: it is sense experience – sight, sound, smell, taste, pain and pleasure. Just how far down the animal hierarchy 'sentience' reaches is a matter for debate: apes certainly have it, fish may have it, spiders don't seem to have much of it, and amoebas almost certainly don't. And just how similar human sentience is to that of other animals is also a matter for debate. The debate can circle round various questions. Is sentience delivered by the sensory systems of different beasts so different as to be scarcely the same thing? Is human sentience always contaminated by the higher-order forms of consciousness that we uniquely have? This question is often presented as a question about whether the content of human experience can ever be 'aconceptual'.[5] Sentience is most readily thought of as sense experience, conceived of as being as innocent as possible of explicit general concepts. Sentience corresponds to the openness of the sensory field around the body. The important point is that merely sentient non-human animals are *lost* or *dissolved into* their sensory fields. They do not experience themselves as embodied subjects located in the field of which they are the implicit centre; they do not have this sense because, as we have said, they do not have the sense of being themselves, period.

Human beings, by contrast, *are* explicitly located in their sensory field. They are explicitly related to the objects within it, which are consequently experienced as being at such-and-such a distance, and in such-and-such a direction from their bodies. Their awareness locates itself among that which it is aware of. I am aware of the objects disclosed to me as being arrayed in a field surrounding me: they are near to and far from me, to the left of or to the right of me, and so on. I spell all this out as a preparation for introducing a contrasting pair of terms: 'indexical' and 'deindexicalized'.

The adjective 'indexical' and the related term 'indexicality' refer to notions which have been explored and elaborated by philosophers of language and of consciousness, by linguists, by specialists in semiotics and a variety of other people. Some of us find these notions are endlessly fascinating but I am conscious that they may be less so to some readers. I shall therefore focus on just one aspect of indexicality: a kind of reflexiveness. When I use the word 'I', the referent of the word is Raymond Tallis, the person who is using it. The word is an indexical in that sense; and this is the sense that I want to borrow when talking about what I regard as the ground floor of distinctively human consciousness: 'indexical awareness'. This is a form of awareness that, as it were, points to itself, or points to its own source. The awareness points back to the one who is aware. I am a touched toucher, a seen seer. In short, I am present to *myself* in the field of things that are present to me. I am next to the object and the object is next to me. The space in which I move is egocentric; that is to say it is centred on my ego, on the embodied subject that I, a human being, am. When I am aware of a physical object, this awareness is of something that confronts me, that I can make use of, must negotiate, or am threatened by, and so on. Neither the object nor myself is dissolved in the space of sentience: we are both located in it, though I alone am aware of this.

So what has this got to do with pointing? It is the necessary condition, the 'deep background' of pointing. Pointing makes the relationship between the embodied subject – present to itself among the objects that are present to it – more explicit. This is especially evident in the most typical mode of pointing: when I point to a physical object that is in front of me. I make myself and the object stand out, along with the relationship of the one to the other. The object and I are italicized, usually for the benefit of a third party. But there are more advanced forms of pointing: we may point *out* many more things than we point *at*; we may draw others' attention to abstractions as well as to concrete things. These abstractions belong to a shared, collective space that goes beyond the scope of indexical awareness, of an awareness that relates to the body of the subject. We may point out that, for example, the weather has been milder; or that someone's son is quite the little gentleman nowadays, and so on. This form of awareness is not merely 'transindexical': it is '*deindexicalized*'. Let me say something about this.

That which is pointed out to us through, say, speech, is only indirectly revealed by the senses. We have direct sensory access only to the sounds that make up the utterance. The referent of speech is not physically related to our body. Surely, it may therefore be thought, fully deindexicalized pointing *out* has nothing to do with pointing *at* in the literal sense, nothing to do with digital pointing. And, admittedly, the origin of the vast realm of deindexicalized awareness – of abstractions, and generalities, and facts by which humans supplement the natural world that surrounds their bodies – is ultimately mysterious. I believe, however, that gestures, and above all pointing, play a crucial role. Pointing is the first step towards the opening up of a realm that is made up of general possibilities, of entities that do not have a physical relation to the body of the embodied subject.

When the producer points to something, the consumer must, as we have said, put himself in her shoes; to assume the viewpoint from which the pointer arises. The pointee is looked for along a line that links it with the producer. The direction of the lines is marked out by the pointing limb. In the simplest case, the pointee is in the consumer's sensory field: I look where you are pointing and I see what you are pointing at. But producers do not confine themselves to pointees that fall within that part of the sensory field they share with consumers. Producers can point beyond the consumer's current sensory horizon and it is this that is crucial to the transition from indexical to deindexicalized awareness.

Reaching out to (General) Possibility

Let us begin with the least complicated case. You are at a vantage point – say, the top of a hill, or halfway up a tree – from which you see something of interest to both of us: a predator, an enemy, a long-awaited friend, food, a beautiful sunset. You point to it in order to draw my attention to it, or even to invite me to join you so that I, too, can see it. At the time you point it out to me, I become aware of the object but aware also that it is outside of my sensory field. I am aware, that is to say, of an object that has no precise location with respect to myself. Pointing opens up, or confirms the prior existence of, an 'over there' populated with objects, that has an indeterminate relationship to my body. What is more, the pointee is also indeterminate. I do not know exactly what you are pointing to. As such, the pointee has some of the essential characteristics of an object of *knowledge*, of the kinds of entities that figure in facts and discourses, which are *general*, with only a proportion of their characteristics fully determined.

First of all, at present it lies beyond any experience I have of it. This affirms the object's status as something that exists independently of my sense experience. In this case, its existence – even though it lies outside of my sensory field – is affirmed by your current experience of it. To some extent, this is already implicit in those instances of pointing where the object actually is in the shared sensory field of producer and consumer, in the interval between the moment you point something out to me and the moment when, after a search, I locate what it is you are pointing out. When you are apprising me of something from a vantage point that discloses things to you that I cannot see, there is a longer interval between my becoming aware of the possibility of the object and my seeing or otherwise experiencing it – while, for example, I move to join you at the top of the slope or tree or whatever. During this time, I more explicitly entertain the notion of an object that exists independently of my experience of it.

This is connected with several other things. Most immediately and intimately, it is connected with the explicit acknowledgement of the public nature of the object and of the world in which it is situated. This public world is sustained by a collective of conscious-nesses, of individuals who have viewpoints different from, but no less real than, my own – indeed, that are at a fundamental level equal to my own. These viewpoints do not merely harvest different aspects of the same objects as those harvested by my indexical awareness. They stand in relation to, or make available, objects of which I am indexically unaware. In this lies the notion of a world that is the joint product, property and arena of a multitude of con-sciousnesses. The as-yet-hidden object to which you are pointing is an object of deindexicalized awareness, an awareness of things 'out there', where the 'out there' has an initially imprecise relation to myself as an embodied subject and ultimately, once we enter the world of words, no spatial relationship of this kind. The world in which the object is located – for example as a referent of speech – is one that belongs uniquely neither to you nor to me but to anyone; a world of which I am not the centre; a world in which 'I' becomes 'anyone'. The 'out there' to which you point is not an 'over there' linked in a privileged way to my 'over here'. To the deindexicalized object of awareness, there corresponds a generalization of myself as *anyone*.[6]

In the interval between my being aware of you pointing some-thing out and my seeing what it is you are pointing at, the pointee has two characteristics that are relevant to the transition to speech and the opening up of the deindexicalized realm: it is indeterminate; and it is only a possibility. These two characteristics are connected: possibilities are of necessity general, just as actualities are of necessity particular (though they may be described in general terms). As I run to the top of the hill from which you are pointing, I am running towards an indeterminate possibility; more precisely, to the (possible) realization of a general possibility in a particular actuality.[7] The status of the pointee as a general possibility underlines its lack of physical relation to myself as an embodied subject: general possibilities do not have physical locations; more precisely, they are not located at a particular distance from my body in a particular direction. Possibilities no more stand in a physical relationship to my body, than they stand in a certain light. It is important to emphasize that the 'de-location' of the, as yet invisible, pointee is more radical than mere incomplete specification of its spatial location and the limitation of my knowledge of its location to the fact that it is 'roughly over there', in the direction you are pointing to, in a space I cannot see. This becomes clear when you think that the possibility might be unrealized – you might be deceiving me or be yourself deceived. There is no way in which a possibility *qua* possibility could be physically related to me. While it is self-evident that unrealized possibilities – things that do not exist – cannot be physically related to my body, this is equally true of possibilities that later turn out to be realized.

The fact that pointing may take place in the absence of the corresponding pointees – in order, for example, to mislead someone – shows the link between pointing and the origin of, and the contrast between, fully developed notions of truth and falsehood. In the material world of actual entities, there is neither truth nor false-hood. 'That which is, is', and that is the end of it, as the early Greek philosopher Parmenides argued so forcefully, drawing from this some profoundly paradoxical, even absurd, conclusions.[8] Truth and falsehood require the entertaining of *possibility* by an individual who has a sense of actuality that goes beyond mere sentience. To a degree, possibility is already bundled in with the objects we en-counter with our sense: objects are an inexhaustible source of as yet unexperienced experiences. The *judgement*, which exceeds what is given in present experience, that an object is of such and such a nature, as a result of which one has certain expectations of it, of the future experiences to which it may give rise, or future uses to which it might be put, lies at the root of the categories of the true and the false. Possibility and the possibility of error, that is, already arise with indexical awareness of objects. Object awareness, unlike mere sentience (sensations of warmth and cold, say), is *corrigible*. Truth and falsehood within indexical awareness, however, remain embry-onic, incompletely explicit. They develop fully only within the common world postulated in the realm of deindexicalized aware-ness. Within that realm, we have not only possibilities attached to actual objects, but possibilities attached to objects that are themselves only possible. These, truly free-floating, possibilities are shared between individuals, in sharp contrast to experiences which are confined to individuals.

Indexical awareness acknowledges the publicity of the objects of which it is aware: it is an awareness that is aware of itself; of its limi-tations; of the fact that the object is being experienced from only one of an unlimited number of perspectives. Nevertheless, this form of awareness is still realized in an individual, perspectival encounter with the object: the intuition of the non-perspectival reality of the object is still had through the lens of perspectival experience. I look at the cup from a particular angle and in a particular light and see that the cup can be seen from different angles and in different lights and that, in itself, it exists at no angle and in no particular light. It is this that is captured in the word 'cup'. By contrast, the as-yet-invisible pointee is, for the consumer who cannot immediately see it, wholly located in a realm which belongs to all: a common world in which no individual is the centre. Admittedly, the objects of deindexicalized awareness may be fitted into a particular history – my history, your history – and in this sense have a personal location, but they are not tethered by spatial relations to the body of the subject; on the contrary, the subject gains access to them through another person, as pure possibilities, free-floating, extricated from the actual. The division, within such possibilities, between those that are realized and those that are not thus cuts deeper; and the distinction between truth and falsehood, and consequently the reality of both truth and falsehood, is sharper. The one who points opens up a world of objective, that is to say 'deperspectivalized', reality, one that goes beyond that of mere material objects like cups and saucers and trees and rocks, for all that the latter are the final touchstone of truth. This is the reality of the world as a network of facts as articulated in language.[9]

Of course, we must assume that, over the last 40,000–100,000 years or so, language has gradually assumed the central role in directing humans' attention to things that lie beyond their sensory fields. One could, however, imagine that even after humans began speaking to each other, there was a transitional phase, analogous to that seen in developing infants,[10] in which the producer positioned on a vantage point might supplement her pointing with single words – 'Tiger!', 'Water!' – that would direct the consumer as to what he might expect. At the earliest stage, these additional indications might be based on primitive calls, corresponding to fear, joy or whatever. I have refrained from emphasizing such auxiliary features too early in our discussion in order to avoid underplaying the absolutely fundamental difference between human and animal consciousness and human and animal communication which provides the necessary background to the emergence of both pointing and (true) human language. The use of language builds on, and presupposes, what has already been established prior to language: even when it is only at the single-symbol stage, it is on the far side of a huge gap separating mankind from the rest of animality. It is this that I have tried to make visible through pointing. The animal that was one day going to speak must first have been so constituted as to be able to point and to build up a world picture that is composed initially of possibilities rooted in objects related to the body; then of possibilities that have only an indeterminate relation to the body; and, finally, of possibilities that have no relation to any particular body and are entirely general. Pointing underlines and expands the bubble of a distinctively human, shared space, that will eventually host, and be massively expanded by, discourse.

One final remark. The index finger can be used to indicate because it is mounted on a body that is a conscious agent rather than a mere organism. But it then helps to develop and elaborate the consciousness of the agent to a higher level, one that is more elaborately shared – more self-conscious and more consciously shared. An aspect of this is the role of pointing in promoting inquiry. As we have already noted, the sense that there is more to be sensed haunts even indexical awareness; but it is extended by pointing. When, from a vantage point, you point to something I cannot see, you take my ignorance-driven gawping, scanning or scrutinizing on to a higher plane. As I travel to find out what you are pointing at, I am engaged in a primitive version of explicit inquiry. Pointing raises questions; pointing at an invisible object puts the ignorance of the knowing consumer to a higher level of explicitness and sends him in pursuit of answers. What is more, pointing at an object that is invisible to me marks a step from pointing *at* to pointing *out*. Pointing *out* has abstract objects. Such abstract pointing – pointing to possibility, to significance, to meaning, separated from material objects – is a fundamental precursor to human discourse, to our peculiar mode of being-together; to our communities that go beyond the spatial cohabitation and rigid interlocking of behaviour, beyond the dove-tailing automaticities of animal groupings that ethologists would like to persuade us are analogous to human societies. We are unique in having a public domain, a place in which truth and falsehood emerge. This domain is first of all stitched together by joint attention made explicit through signs, foremost among which is the gesture of the index finger.

There is more to be said about the metaphysics of pointing but now let me instead turn to the question of whether animals truly point. If I am right in arguing that pointing grows out of something fundamentally and distinctively human, we must presume that they do not. Let us examine the evidence.

chapter three
Do Animals Get the Point?

One of the most cherished assumptions of
contemporary psychology: namely that ape minds
and human minds are in fact basically of the same
type and shape, that there is no great qualitative gulf
between human ways of construing the world and
apes' ways, that apes are in effect just like us, only
less so.[1]

The previous chapters may have made some readers cross. Pet-lovers and many animal biologists will be unhappy at the confidence with which I have set aside certain claims that are often made about the capacities of animals. Once it has been made clear what indexical and deindexicalized awareness require, and what are the prerequisites for participation in the pointing game, and how these are ultimately dependent on unique features of *H. sapiens*, one might expect that even the most fervent animal-lover would be prepared to concede some ground. But arguments may be irrelevant, if they have so much interest in attributing distinctively human cognitive characteristics to animals. While pet-lovers may feel a need to justify their emotional investment in the beast who seems to share their life, evolutionary psychologists and their kin want to justify a Darwinosis – a pathological variant of Darwinian thinking – that promises to make what they have discovered about animals relevant to the politics, sociology and psychology of human life.[2] Both parties seek acknowledgement: the former from their pets, for which it will be necessary to believe that their pets are more like humans than dispassionate observation would suggest; and the latter from the community to which their writings are addressed.[3]

This chapter, which will make me no friends, is designed to address head-on the issue of whether or not animals point. They do not, I will argue, because they do not have the cognitive wherewithal, and the correlative world-picture, described in the previous chapter, which would enable them to get the point of pointing. Their failure in this regard is an important indicator of the gulf between us and them.

Concerning the Dog B.

Let me begin with a personal experience. For some years we were the proud, if at times exasperated, owners of a handsome flat-coat retriever, whom I shall call 'B.'* Though he was an extremely good-natured beast, B. used to wear us down with his obsession with the eponymous act of retrieving. From brainrise to brainset, he would present us with a ball or a stick or some other projectile to throw in order that he could chase after it and return it to us. Since he was a retriever, it might be thought that we had no grounds for complaint: the sun shines, spiders spin and retrievers retrieve, that's how the world is. However, he was not very good at retrieving, which is sur-

---

* His name was pronounced 'Barkleigh'. I thought he was named for the philosopher Bishop Berkeley, who argued that objects existed only insofar as they were being perceived or were themselves perceivers. My children, however, thought he was named for Charles Barkley, the great basketball player, whom they worshipped.

prising, given that this was what he did, or wanted to do, all his waking hours: a case of endless practice not making perfect. The reason he lacked the requisite skill was quite straightforward. Because he tended to look at us rather than the missile when we threw the latter, he would often fail to see where it went. Not infrequently, we would have to retrieve the object ourselves – an irritating variant on the situation of 'having a dog and barking yourself'. Significantly, pointing in the direction of the object was of no use. B. was unable to understand where one was pointing or, indeed, *that* one was pointing. He had no grasp of the referential nature of pointing.

At the outset, I noted how pointing, apparently the most transparent and natural of signals, is nothing of the kind. Wittgenstein, with the Martian eye of a person with Asperger's syndrome, saw that it is not self-evident, when one is pointing, that the object should lie along the line connecting the shoulder with the tip of the index finger, nor that the direction indicated by pointing should be shoulder-to-finger rather than finger-to-shoulder. It certainly wasn't evident to the dog B. And the problem goes deeper than unresolved ambiguity as to which way a pointer is pointing. The direction of pointing is only self-evident if you have the *idea* of pointing, more specifically, the idea of someone wanting to point something out to you – the idea, in short, of meant meaning mediated by a sign.

The deeper reasons for B's failure in this regard will be apparent from the discussion in the previous chapter. B, being a dog, not a human, had a consciousness that had not woken out of sentience. He had no sense of objects independent of himself: lacking the existential intuition 'That I am this', he did not experience himself as an embodied subject. Nor was he capable of having the sense of another embodied subject which might have information that he himself did not have. Nor, finally, was he able to understand the rules governing pointing – the relationship between the pointer, the pointing finger and the pointee.

Pet-lovers and others might protest that, even if B. were not up to the mark, there are dogs – pointers – who do seem up to pointing. Indeed, the *OED* gives an eleventh meaning of 'to point': 'Of a hound: To indicate the presence and position of (game) by standing rigidly looking towards it.' And this was what B. used to do with a missile. He would drop it at one's feet and stare at it fixedly. This, surely, was an example of pointing, not withstanding that it was his directed gaze, underlined or amplified by his good-natured, handsome head, that was the pointer. (If he had used one of his legs to point as humans do, that is, raised right up, of course he would have fallen over, so he may be forgiven the non-canonical form his apparent pointing took.) There are, however, several reasons other than those I have already mentioned for denying B. and his cousins the pointers the credit for having this one shot in his locker.

First of all, one should be suspicious of one-trick ponies. If pointers truly point at game, or game-equivalent, such as a thrown-and-retrieved ball or stick, by looking at it, why do they not point at other things and under other circumstances? They do not point by other means (for example using other parts of their body) or to other kinds of objects. This shows at once they don't grasp the underlying principles of the pointing convention: their pointing is acquired by dumb imitation or wired in by instinct. Secondly, the producer (the dog) and its pointer (in this case its head) are not clearly differentiated. In human pointing, a part of the body, but one remote from the centre of consciousness (the latter usually under such circumstances felt to be behind the eyes), the hand is utilized by the embodied subject. When we use our finger or arm in the indicative mode, it is the tool of our agency. The stance assumed by the (canine) pointer does not have a differentiated producer and pointer. The whole body is involved; or, more precisely, the part that is involved is not clearly differentiated from the part that is not. The dog, that is to say, does not *see* itself orientating its body to the target object; whereas in humans the index finger, and the targeted object and the relationship between them, are all gathered together in the visual field and the consciousness of the one who is pointing. You can see yourself making the link with the object, and you have a clear idea of its visibility to another person. Thirdly, canine pointers are remarkably obtuse when it comes to understanding others' pointing, which was where we began. However hard we tried to communicate what we meant, B. failed to look in the direction we were pointing. The proof of the pointing, as *meant meaning*, is in the consuming. If someone truly understands the pointing game, they will consume indicative gestures as well as producing them, just as a true speaker is also a comprehending listener. In infants, comprehension of pointing precedes production of the gestures. This observation is a cue for us to move from theory and casual observation to experiment.

Concerning Apes

That dogs do not truly point or understand pointing, is hardly surprising, given that not even the great apes, our closest animal kin, have the capacity to participate in referential pointing. The primatologist Daniel Povinelli has investigated the widely accepted belief that chimpanzees, the most intelligent of the apes, can in fact do this.[4] His experiments clearly show that they do not, notwithstanding that they deploy gestures, such as holding out a hand to beg for food, that structurally resemble pointing. This latter gesture is not a generalized indicating or referencing device. 'Gimme!' is not a sharing of information.[5] What is more, chimpanzees do not point in the wild. They do, however, seem to point in captivity, although only for their human carers, not for each other.[6] Has living with humans brought out a latent understanding of the referential nature of pointing, the idea of meant meaning, and the notion that others have internal mental states?

Carefully controlled studies have yielded an unequivocal 'No'. In Povinelli's key experiment, the experimenter points to a row of boxes, only one of which contains a reward. The chimpanzee will select that box if the experimenter's index finger is closest to the box containing the reward. If, however, another box is closer to the index finger, even if the finger is not pointing at it, the chimpanzee will select that one – an incorrect box. When the pointing finger is equidistant from two boxes, but none the less clearly references only one of the boxes, the apes choose boxes at random. The chimpanzees rely on the distance between the finger and the box and not on the referential significance of the finger to decide which box to go for, and they identify the pointee as whatever object is closest to the finger, irrespective of the direction in which the finger is pointing. In summary, as Povinelli concludes, pointing – rarely produced, and then only in captivity – is not a referential gesture but 'a local physical cue to the location of a reward'.

Two-year-old children, by contrast, understand pointing, even under the most difficult circumstances, and choose the box referred to by the tip of the finger, irrespective of the distance of the finger from the box. Unlike chimps, they get the point. What is more, pointing at an invisible object does not prompt a chimp to search for something corresponding to a pointee, as it does in humans. If chimps are not capable of pointing in an understanding way, it is hardly likely that dogs would be capable of this. Our ascription of pointing to pointers is simply another example of how we project into beasts our own mental capacities and world pictures.

Povinelli's experiments on pointing are part of his wider inquiry into the way chimpanzees understand the world – how it is put together, what it is made of, and why it works the way it does – and he dismantles the seductive assumption that similarity of behaviour means similarity of mental processes. A series of ingenious studies demonstrates just how remote the chimpanzee mind is from the human mind. For example, while one chimp may follow another chimp's gaze, thereby seeing what his fellow is seeing, this hard-wired response does not require the first chimp to have the idea that the other chimp has a mental state of seeing. This deficiency is revealed by experiments that demonstrate that chimpanzees do not have the notion of another – say a human observer – being unable to see something because their eyes are covered. Contrast human infants, who are able to appreciate pointing, and the notion of seeing as a state in another person, before they reach their first birth-day. What is more, they produce pointing as well as understanding it, and they point declaratively (to share information) as well imper-atively (as a digital equivalent of 'Gimme!').

At the heart of the chimp's failure is an inability to grasp what Povinelli calls the 'proto-referential' nature of pointing. He links this with the inability, also evident in all other animals, *to form the notion of another creature's having a mind*. Without such a notion, it would not be possible to grasp the idea of *meant meaning*. Equally, as we have already observed, it would not be possible to appreciate that others have different viewpoints that are advantaged or disadvantaged compared with ours with respect to some piece of knowledge. This is a prerequisite of participating in the pointing game.[7]

The fact that major long-term studies of chimpanzees in the wild, some of forty years' duration, have reported no evidence of pointing,[8] notwithstanding its apparent simplicity and utility, is entirely consistent with the animals' failure to understand pointing. Indeed, if you do not do pointing, it seems unlikely that you would be able to make sense of it. Not pointing things out to each other is also consistent with something else: chimpanzees rarely if ever teach others, even their young. The latter acquire skills by observation and by trial and error, not by training. Animals that do not point do not point things out to each other in a broader sense.

Pointing as Proto-Reference

I want to look a little harder at the notion that pointing is 'proto-referential'. This is intuitively attractive but we must be careful, when we succumb to this notion, to emphasize the importance of the prefix 'proto-'. Just how 'proto' pointing is as reference is obscured by the fact that, for most of our lives as consumers and producers of pointing, we are also in command of language. Pointing therefore takes place against a background of, and rides on the back of, verbal discourse. This, as we shall see when we discuss 'ostension' in Chapter 5, can lead us to over-estimate what is achieved by silent, or grunt-accompanied, pointing on its own. We imagine not only that we pick out, but, in some primitive way, refer to, an object by pointing alone.

There is an important truth, however, in the exaggerated claims about the referential power of pointing. Pointing connects us with an object at a distance and this connection is mediated by a sign, the extended index finger. The sign, for all that it is in part conventional, is not entirely arbitrary: it is also iconic or pictorial. The finger is used to instantiate, fatten and make explicit the axis of the line linking the producer with the pointee. This indirect targeting of the attention of one person by the action of another has some of the characteristics of reference and the pointee consequently has some of the characteristics of a referent.[9] If to 're-fer' is to 'bring back' something in a virtual way, making it available to another in a mode that is separate from immediate sensation, pointing certainly does that. The object is taken hold of as the meaning of a sign, and thus it can be delivered to another person by re-directing their attention. It is 'holding close at a distance', bringing it nearer to someone by moving their attention towards it.

We must not, however, ascribe full referential status to pointing. The term 'reference' 'has to do with the relationship which holds between an expression and what that expression stands for on par-ticular occasions of its use'.[10] The pointing stands only for the line that is realized by the pointer, as its axis. Without the direct visual attention of the consumer, the pointee is not picked out and does not have the status as a referent. Words work without glances – indeed, they work without the directed bodily attention of the person to whom they addressed. The realm within which words characteristically operate – their home territory as it were – is not the sensory field surrounding the body; it is, as we have already noted, a deindexicalized realm whose places are not typically located with respect to the body of the speaker (or writer).

There are situations in which pointing may seem to be *predicative*, as Herbert Clark has argued.[11] Supposing I ask 'Where is the cat?' and you respond by pointing to it, you seem to be conveying a predicate, roughly translatable as '— is over there'. But this is an illusion, because, just as reference cannot exist without some actual or implicit, uttered or silent, predication, so predication cannot exist without a referent to which it is attached. The seemingly predicational property of the pointee is due entirely to the framing that has taken place in words. Silent pointing would not deliver a predicate, even less a full assertion such as 'The cat is over there.'

The very fact that pointing is, to use the linguists' term, 'holophrastic' – in other words, all of the seeming grammatical components are fused in or bundled up in a single sign – means that it is not correct to differentiate referring and predicative functions within it. Without the aid of language, it has neither of these functions. Nor can pointing on its own be used to make an assertion: it is not a 'telescoped' assertion. If the producer points to an object that the consumer sees, so that the latter sees that the object exists, it might be argued not only that the consumer's attention is drawn to the object, but also that the pointee's existence and/or location is made explicit. That this is not quite the same as asserting the existence of the pointee becomes clear if we try to imagine an *untruth* being conveyed in silent pointing. This is simply not possible without prior verbal discourse. We cannot, without the assistance of language, point to an object that is not there in order to tell a lie in Pointish. It is, in other words, not possible to make stand-alone assertions that can either turn out to be, or not to be, the case. A full-blown assertion has to be something that can be false. If it cannot be false, it cannot be true either.

Back to Beasts

This highlights another connection: that between pointing and the invisible world. As already noted, chimpanzees do not have a sense of others' experiences: they do not ascribe the state of seeing to others. More generally, they do not have a theory that others have minds. Minds are, of course, hidden: we become aware of them only inferentially, and humans alone make such inferences. Humans, who are upright creatures, for whom vision is the dominant sense, and who experience themselves as embodied subjects, and a world as being made up of objects that have hidden properties and connections, feel constantly encircled by the hidden. Pointing puts the hidden, that which is about to be revealed, into italics. It highlights both what is revealed and what is concealed. It does this most obviously when we point to hidden objects by pointing to where they are hidden, or when we point to a place, say an empty chair vacated by the pointee. We can do this, of course, often only with the aid of language. Animals do not have this sense of the invisible – whether it is that which is hidden within objects or the causal links between them or the psychological states hidden within others in virtue of which they are seeing, noticing, or in other ways aware. As Povinelli says, chimpanzees' insights are determined by the situation they see in front of them in the literal optical sense. What they see is what they cognitively get.

We shall return to the invisible world in the final chapter. For the present, we note that it is easy to see how full-blown human pointing contains inchoate stirrings in the direction of the capacity to make reference and to assert *that* something is the case; of reference to an object or state of affairs and assertion *that* it exists or is the case. It does not, however, get all the way to either reference or predication. And this makes it easier to understand why no animal, not even our nearest and brightest kin, the apes, makes use of pointing. We have seen how the proto-referential import of pointing escapes them entirely. It is too tempting to imagine that when beasts call to each other, they are effectively 'referring' to objects and states of affairs that those calls are 'about', and so 'make assertions' about them. By this means, it is possible to elide the huge difference between beast and man. It will already be evident, particularly from what we have said in Chapter 2, that animal calls, which attract the attention of other animals and may prompt certain behavioural responses, do not count even as proto-assertions. They do not originate from producers blessed with either indexical or deindexicalized awareness. There is no gap between animals and their field of sentience: they are not embodied subjects located explicitly within their sensory fields. 'Reference', which presupposes that gap, reaching across an explicit distance, is not, therefore, possible. The gap in linguistic reference is the gap between indexical sounds that are located with relation to the body of the speaker and the deindexicalized realm of possibilities accessed through generalities. True pointing is halfway to this gap crossed by reference.

Assertion is equally impossible for animals. To assert *that* something is the case can take place only in the context of a nexus of entertained possibilities, some of which may be, and some of which may not be, realized, only in the context of differentiated truth and falsehood. Such a context does not arise for animals for whom the world is what they currently experience, who do not have a sense of objects beyond what they sense, other than themselves. Only for a creature living in a world that is a nexus of (realized and unrealized) possibility can verbal or non-verbal signals become assertions, communicating what is the case. Only for such a creature can what is the case be asserted; and can what actually exists be assertible as what is the case. Only such a creature can harbour the intuition *that* it is the case. We may think of real pointing as '*Ur-that*', as the beginning of the collectivization of 'that', an explicit public realm which is collectively acknowledged in the community of minds that humans uniquely sustain. With pointing, as we have noted, we move from the indexical to that deindexicalized realm in which we spend most of our conscious existence, related – in joy, fear, hope, intention, etc. – to things we cannot see but can envisage.

This leads us to consider the whole question of animal communication by means of bodily exhibition. Some might suggest that, while beasts might not in theory point because, ultimately, they do not have a sense of themselves as embodied subjects, in practice they use their bodies to communicate in a way that is susceptible to as complex an analysis as pointing. Animals grimace, blow themselves up, present their rear ends, etc. in order to communicate something of their states of mind to other beasts to influence the latter's states of mind. We can now see that while it is possible to cook up a tendentious description of such behaviour that involves self-conscious creatures trying to manipulate the consciousness of other creatures whose minds they are conscious of, the utterly standardized, non-creative forms of these 'communications' and the fact that they are never elaborated and scarcely evolve from generation to generation, provide overwhelming evidence that they are mediated merely by 'generalized action sequences' embedded in sensorimotor systems, rather than being guided by higher-level intuitions of themselves and of others. That is all that is possible before the founding intuition of one's self as an embodied subject surrounded by incompletely known objects.

The limitations of non-human creatures, even of our nearest animal kin, have been beautifully demonstrated by Povinelli's experiments. As he concludes, chimpanzees do not 'share the [human] ability to conceive of abstract unobservables as explanations of the social and physical events that cascade around them'. This inability on the chimps' part is equally evident whether we are concerned with social behaviour, where the variables 'take the form of mental states, such as desires and beliefs', or whether we are concerned with tool use and manufacture, where the relevant variables are 'phenomena such as gravity, force, mass and the like'.[12]

Let me end by returning to the good dog B. When we see him looking at his ball on the ground as a way of drawing our attention to it, it is tempting to unpack from this behaviour a rather impressively complex tangle of relations: the dog being aware of me, and of my insufficient awareness of the ball, and of the need to raise my awareness by making explicit the ball through drawing attention to the line linking itself to the ball, so that said awareness will be directed down towards the ball. The reason we should resist this temptation is that, as we have said, the creature is a one-trick pony. If the dog were truly pointing – with all the mental capacities this implies – we would expect it to exhibit a much wider and richer range of communicative behaviour that it mobilizes in its dealing with ourselves and other dogs. If the dog's 'pointing' were really informed by these capacities, then we would have spent less time when B. was alive retrieving the sticks ourselves.

chapter four
People Who Don't Point

Learning to Point

When our first-born was about ten months old, and before he had any significant language, he started pointing at things. My intense pleasure and astonishment at this event took me by surprise. In retrospect, my reaction is less surprising. Parental joy at seeing our complex common world being constructed in the mind of a new-comer to the planet – in a mere handful of months, and out of quite inadequate materials hurled somewhat at random – is well founded. Pointing is on a par with the first exchanged glance, the first smile that is not attributable to gut sensations and is seemingly genuinely addressed to its recipient, and with the first word. It is another reminder that someone is developing 'in there' and that that someone senses that you are 'out there'. My pleasure, then, was more justified than the customary projected egocentricity of the adoring parent.

It will be clear by now why pointing is an important marker of child development: it reveals so much about the way in which a child is relating to the human world. We have talked about some of these things already, the most obvious of which is the intuition of another person seeing the world from a different angle and so, possibly, having a cognitive lacuna which you can fill. Behind this is the idea of the lacuna being filled by a communication between one person and another. And, behind this again, the even more recherché notion of the material in which the lacuna is located: the other person's mind. Pointing, the invitation to joint visual attention, is a sign of an emerging mode of togetherness. The blob in the cot has taken a giant step to becoming an interlocutor, a companion of one's days.

Little wonder, then, that the emergence of pointing had such an impact on me. Sometimes, it made me laugh. It still does now. When I see a child, its index finger at the ready, cocked for ostension, in *status demonstrandus* as it were, it seems as it sits in its pram to be in a permanent state of preparation for pronouncement. The raised index finger commands attention and prefigures future assertions of control – Harken! – and authority. At any rate, once it is established, it becomes the dominant gesture. By a child's first birthday, pointing accounts for more than 60 per cent of gestures.

Infant and early toddler pointing has many facets. The child may point to something that has attracted its attention on account of its beauty, its interest, its unfamiliarity, in order to share it with someone else. In pointing *at*, it points *out*. Alternatively, it may point to an object in order to find out what it is called. Such pointing may be supplemented by 'Wazzat?' to indicate that the communication is a question rather than a proto-reference, or a question fastened to its target by a non-linguistic quasi-reference: pointing is used to specify the entity whose name is sought.[1] Pointing may also be used by the child to show that he knows the meaning of a word, as when a parent asks him to point to 'the doggy', 'the tree' and so on. It is because it says so much about the triangular relationship between the child, her language and the community into which she is being inducted, that it is so fascinating.

Infant pointing has occasioned an extensive scientific literature, in which psychologists try to work out what is embedded in this gesture made by a ten-month-old with an unimaginably different consciousness. In a classic paper published over thirty years ago, Bates and others described how infant pointing could be either 'imperative' or 'declarative'.[2] In imperative pointing, the infant, with all the arrogance and unbounded egocentricity of a creature whose ego has not fully formed and hence located itself in a world of which it is, to its chagrin, not the only occupant or even the centre, will obtain certain objects by exploiting others – usually its careworn parents – as social tools. This is primarily manipulative (and being manipulated by one finger only is the condition of being a besotted parent). It is only secondarily, if at all, communicative. It might not even require attribution of mental states to the downtrodden social tools pushing the pram or rocking the cradle. Declarative pointing, by contrast, is 'grounded in an understanding of its communicative function…entailing an understanding of others as mental agents whose mental state can be influenced through the pointing behaviour'.[3] This mode of pointing is somewhat more collegial, with the infant perhaps wanting to inform his interlocutors or to share an experience so that emotions might be aligned, rather as we point things out to each other in the spirit of togetherness. Tomasello and others have recently built on this dichotomy and attribute to the infant 'not only an understanding of others as mental agents but a sophisticated use of referential communication and highly social and uniquely human motives'.[4] Needless to say, psychologists being what they are, and the mind being such slippery stuff, this interpretation has been challenged by others, such as Southgate and colleagues, who question whether infants ever point altruistically. The children fuse imperative and declarative modes of pointing into a self-interested 'interrogative' or 'information-requestive' mode.[5] The infant, at any rate, is concerned primarily to get the adult to do something for it rather than to inform the adult or to share an experience. Others have observed that imperative pointing never precedes declarative pointing, prompting George Butterworth to comment that the understanding of others as 'agents of contemplation' grows in parallel with understanding them as 'agents of action', of tools to bring things about.[6]

Failing to Point

Whether or not infants point in order to inform others or to share experiences, demonstrating collegiality even while they are still filling a nappy, there is a consensus among developmental child psychologists about the importance of becoming a pointer. The failure of this neurodevelopmental landmark, particularly in the context of apparently otherwise normal physical and mental development, may be an ominous sign, and non-pointing children may go on to develop the features of autism. Even in such cases, the failure of pointing is not complete. Here is an account of two autistic children, George and Sam, by their mother, Charlotte Moore:

> Both George and Sam did point, and at the right age, too…
> but I think the difference is that they mainly pointed on
> request, 'Where's the moon?' Point – 'Well done, darling!'

They also pointed to get their needs met. Biscuit tin – point – result. But what they didn't use, or used to only a limited extent was the shared-attention point. They didn't call my attention to things in order to make me see, enthuse, inform or share.

As with so many developmental details, I didn't fully appreciate the difference until I had Jake [her third child, who did not have autism]. By ten months, Jake's arm was permanently extended in a point. If he spotted something interesting, he had to make sure that everyone else in the room had spotted it too.[7]

For a variety of reasons, autism – a spectrum of disorders with a variety of manifestations and different degrees of severity – has attracted huge attention in recent years. My treatment of it here, subordinated to my overall theme, will be a mere glance. The many impairments of imagination, communication and social interaction that are seen in autism result in what *Webster's Dictionary* describes as 'absorption in self-centred subjective mental activity (as day-dreams, fantasies, delusions, hallucinations) esp. when accompanied by marked withdrawal from reality'. The reality from which autists withdraw is the shared reality of a communal world that includes other people with minds like themselves, viewpoints different from their own, material needs like theirs and emotional needs they can hardly guess at. People with autism may make little eye-contact with others, either because they lack the appetite for reciprocated acknowledgement or because they actively dislike being looked at. The abnormalities of language are the most prominent: some autists have no spoken language at all and some use language in a most peculiar way, for example, repeating the same words, reduced to their sounds, again and again. Conversely, high-functioning autists may rely too heavily on the spoken words, spelling things out unnecessarily and sometimes 'talking like a dictionary'. More often their language is entirely functional, never merely for the sake of communication or verbal togetherness, and excessively concrete. They have difficulties with personal pronouns, particular the first-person pronoun.

The losses are not always global. There are islands of perfectly preserved function. Sometimes, in a small minority of autists who tend to be over-represented in the public mind, there are extra-ordinary talents, such as astonishing verbal or visual memory, a gift for calculus or perfect pitch, but these gifts rarely in any sense compensate for the deficits that make life so difficult for them and those who care for them. They are not connected with wider skills that may be of use in life. The preserved capacities and the occasional extraordinary aptitudes resemble, Charlotte Moore says, 'an archipelago of islands scattered across a sea of confusion. Most islands are not even within hailing distance of each other'.[8]

One of the most striking features of autism is the lack of any sense of a personal, long-term future, of a life to be led long-term as opposed to be lived from moment to moment. Autistic children, Moore notes, have 'no ambition, no thoughts about their adult selves, no perspective on anything except the here and now'. And this, along with the special difficulty with personal pronouns and acknowledging the subjectivity of others – their equal, continuing reality – demonstrates that although autists are self-centred, the self upon which they appear centred is poorly developed. Which brings us to the heart of the matter: they are fastened to what is 'there' in the most literal sense: to what is simply visibly before them. Their 'experiences are disparate, unconnected, a series of sensory impressions that don't add up to make a pattern'.[9] This has been encapsulated by the neuropsychologist Uta Frith as 'a lack of central coherence in cognitive processing'.[10]

Ordinarily, we effortlessly organize the stream of incoming information into more abstract levels of meaning. The absence of this in very small children is often striking: they focus on the incidental details and miss the over-arching point. When our younger child arrived in Cornwall for a holiday after a 300-mile journey, his toddler attention was attracted not by the sea or the cliffs or even by the cottage, but by a slug on a stone. Missing the over-arching point is also evident in small children who are out on a treat. How often one sees parents dragging their child towards, for example, Thomas the Tank Engine, while the child is pointing backwards at some small detail that has attracted his attention. In the case of autists, this persists into adult life.

This impairment of the capacity to incorporate the flow of experiences into a large pattern or frame of meaning or event or timetable may account for some of the special talents that a minority of autists have. They are able to focus on perceptual details with less conceptual intrusion. Those concepts that help us non-autists to make more coherent sense of what we experience also distract us from what we are gazing at: we look and think, and, thinking, cease to see. Intrusive abstractions dust over the windows of perception. For autists, what is there before their eyes (or ears) is less immediately or completely swept up into a schema of what is not there. To put it very crudely, percepts are less sheathed in concepts.[11] According to Frith, autists do not favour globally processed, conceptually condensed interpretations of what they experience, and there is increasing evidence in support of this theory.[12] Autists are better at locating a small element within a whole picture (the so-called 'embedded-figure test'). They include, as we have noted, an unusually high number of people with perfect pitch. They also perform exceptionally well on block design tests that do not require integration across the blocks. All of these observations are consistent with the over-emphasis on detail at the expense of higher-level organization and meaning: they are fine-tuned to detecting small-scale perceptual patterns and associative regularities without interpreting them within a coherent explanatory framework. An autist's experience will not only be unsheathed, their consciousness more naked, but their lives will seem to be 'one damn thing after another'.

Why does this happen? Moore speaks of her own autistic children's 'disjointed sense of their own bodies'. She cites an autobiography of an autist, Donna, who reports that sometimes she thinks somebody has touched her arm and realizes that it is her own hand that has done the touching.[13] One would expect such a disjointed corporeal sense to be reflected in a higher-level disjointedness. As we discussed in Chapter 2, the sense of the coherence and continuity of our own bodies is the foundation for the intuition of a coherent and continuing self extended over time. The sense that 'I am' begins with the sense 'That I am this', where 'this' is one's body.[14] The question then arises as to how it might be that autistic children, who may in some cases be cognitively well developed, and certainly not void of proprioceptive sensations (those sensations that mediate bodily awareness), fail to develop this ground-floor bodily sense of self.

Charlotte Moore observes that 'Babies of three to four months are supposed to play, first with their fingers, then with their toes. They spread their hands in front of their faces and wiggle their fingers; they lie on their backs, catch their feet, and bring them up to their mouths. Jake did this at the right time...George and Sam didn't.'[15]

Which brings us back to the hand and pointing. Autists do not seem to have this combination of bodily discovery and self-manipulation – especially through the master-manipulator the hand – that lies at the root of the fundamental human sense of being an embodied subject, of one who is and who is not, his own body. This may be why they have a less pronounced sense of the objectivity of objects, of objects as being an explicit part of something that lies beyond sentience. This might be expected to result in a limited capacity to tap into a field of meaning that goes beyond the sensory field, which, as we noted, is both presupposed in and developed by pointing. Pointing, as we saw, points to two kinds of beyond: the otherness of the world and the presence of other minds in the world.

This explicit sense of the beyond is absent in all non-human animals, even our nearest animal cousins, the chimpanzees. It is interesting that research has shown that very young infants will turn in the direction in which they see someone else gazing. According to the findings summarized by Daniel Povinelli, by eighteen months children will follow an adult's gaze into space outside their own visual field, precisely locate the target of that gaze and reliably follow a gaze in response to eye movements alone, without any movement of the head.[16] It has been suggested that such behaviour indicates that the child knows that the mother 'is looking *at* something, that she *sees* something, or that something has engaged her *attention*. This is controversial, but it has been postulated as a precursor to more explicit forms of attention-sharing or 'joint visual attention' – what pointing both requires and tries to bring about – or representation of attention in toddlers. Chimpanzees also show gaze-following but they do not understand the difference between targeted and random gaze, between one that is charged with attention to an object and one that is not. Their gaze-following behaviour is cue-driven and they 'do not understand how gaze is related to subjective states of attention'.[17] Autists do not make or solicit eye-contact: the sense of the other person's mind, accessible through their gaze, is poorly developed; and the notion of the other's gaze as a guide, a pointer, not only to what they themselves are concentrating on, to what is of interest to them, but also to what may be of interest to another person, is correspondingly weak, precisely as one might expect in someone who has an attenuated sense of the self and the self of the other person. That children who point in the normal way also increasingly check whether they are being observed and whether their pointing is understood, underlines how what is sought truly is joint visual attention.

It is easy to understand how the creation of an explicit, visible image of attention such as pointing needs to be rooted in something more primitive, less conventionalized, and part of ordinary behaviour. The child who values its mother's gaze, not only in wanting to be gazed at, but also in wanting to gaze at what the gaze is seeing, to be united with the mother in 'joint visual attention', is ripe for communicative pointing using the index finger. Without this preliminary sense, pointing will not occur. The tragic example of the autist, who does not seem to value gaze, who has less awareness of his own outlying parts (his arms and fingers and toes) as being *his*, as his agents, as his signs, and consequently does not point, or point out, or point for, shows how deeply rooted in the nature of human being pointing is.[18]

chapter five

Pinning Language to the World

The Meaning of Meaning

When I wrote the first draft of this book over thirty years ago, I was motivated by quite different concerns. My interest in philosophy had focused on trying to understand the relationship between language and the world; more particularly, how it was that words could be about the world; more specifically still, what 'the meaning of meaning' was and how it was achieved. I was not alone in this, for something rather strange had happened to philosophy in the English-speaking world in the twentieth century.

Under the influence of geniuses such as Gottlob Frege and Ludwig Wittgenstein, the discipline – admittedly not for the first time in its history – took a 'linguistic turn'. There were many reasons for this, but one was the suspicion that philosophical problems often arose out of the misuse of language. We wouldn't be scratching our heads over the nature of 'time' or whether or not we *really* experienced the world as it was, it was argued, if we reflected on the proper use of words such as 'time' and 'reality'. The consequences of this so-called linguistic turn were often dire: thousands of articles and scores of books were devoted to making fine distinctions between different uses of words. But some philosophers were spurred to do truly interesting work, directed not only at developing a philosophical grammar that would clarify the legitimate and illegitimate uses of keys terms in philosophical discourse, but also at exploring, at a fundamental level, the nature of the relationship between language and the world. Its goal was to develop a comprehensive, fundamental and transparent theory of (linguistic) meaning: how do expressions and larger units such as sentences come to have the meanings they have? How do they do the job they do? How can language reach out to, and capture, by expression or representation, extra-linguistic reality?

Needless to say, this project did not achieve completion. Indeed, over a century after Frege, directly and through his influence on Wittgenstein, turned philosophy in the direction of language, answers to these fundamental questions look even more distant, and philosophy has undergone something of an anti-linguistic turn. Part of the reason for this is the philosophical habit of getting stuck on preliminaries. One such preliminary has been trying to arrive at a clear notion of the meaning of 'meaning', and much heated air has been exhaled from many thoughtful heads on this matter. My own contribution, the three unpublished volumes entitled *Of Expression and Representation*, is still bending the floor in my loft, and long may it remain there. Nevertheless, the nature of verbal meaning remains relevant. For although we make sense of the world through words, how we do so is deeply mysterious, and of the greatest possible interest. How we do this mysterious thing promises to cast light on our sense-making consciousness, on the community of minds of which we are a part, and on the nature of the reality into which we are thrown. As for pointing, it seems to promise a means of probing these questions, given that it has often been seen as the tool by which the meanings of words are conveyed to those who cannot speak a language – in short as the key to the infant's passage from speechlessness to speech.

To examine this claim, that pointing can enable us to access the meanings of words by the back door, we need to think about the meanings of meaning, of that in virtue of which words convey meant meanings. This reaches towards one of the greatest and most wonderful capacities of human beings: how it is that, alone of all living creatures, we express the world we live in, and by this means transform it and make it more amenable to our will; how it is that we make sense of the world, rather than merely experiencing it through our senses; how it is that by means of hot air exhaled through our mouths we inflate that massive realm of possibilities and asserted actualities that is the human world, the joint product of millions, nay billions, of human minds over tens, perhaps hundreds, of thousands of years. Pointing seems to offer an alternative route to this linguistic grasping of what is there.

It is easy to think of pointing as being nothing more than (to use the phrase of the nineteenth-century German psychologist Wilhelm Wundt) 'an abbreviated grasp movement'.[1] It is as if it occupies a middle position between physical 'prehension' and conceptual apprehension or comprehension. It mediates between the physical space of the body surrounded by objects and the semantic space of speech.[2] In pointing we organize space around ourselves in a way unknown to the physical or biological reality: we appropriate space for shared humanity, for 'we-being'. Pointing is a halfway house between physical space and the linguistically articulated space of the human world. But before we run with the idea that we can point out, or point to, the meanings of words, we need to get a clearer idea of what we mean by the meaning of a word.[3] And once we have a clearer idea, we may no longer believe, as I did in 1973, that 'Pointish' may be a missing link between the physical field of material objects and the semantic fields of words. Or not in a simple sense, anyway, though enough of the idea stands up to make it interesting.

What is the meaning of a word? What do we mean by its meaning? It is a question that invites easy answers. It is very tempting to think of the meaning of a word as the object, or kind of object, which it names. This so-called bearer theory of meaning has few supporters and it is often dismissed contemptuously as the '"Fido"– Fido' theory: the meaning of the word 'Fido' is the dog called Fido. That is how we can point to the meaning of words: we point to Fido and by this means indicate the meaning of 'Fido'. This notion quickly runs into problems. Are we sure it makes sense to say that the meaning of the word 'Fido' is itself hairy and barks? It should be, if it is identical with Fido, who is hairy and barks. What if every dog called Fido dies? Does the word 'Fido' then cease to have meaning? What is more, not every word is a name, even less is every word a proper name like 'Fido'. What would be the bearer, outside of language, of a general term, such as 'dog', given that all actual objects are particular, or of an abstract term such as 'economic trends', or so-called connectives like 'and' or function terms such as 'the'?

Perhaps the meaning of a word is a mental image. This would certainly cope with the problem of Fido's death. But it has difficulties of its own. For a start, it makes the meaning of words rather unstable. The image corresponding to 'Fido' in your mind, and the image in my mind – or mine on one occasion as opposed to another – might be totally different. Our words would therefore have different meanings, and communication would break down. What is more, it does not make sense in the case of many words to think of them as having a corresponding image. What is the image corresponding to 'economic trends' or to 'the'? If we substitute 'concept' for 'image', the problem of instability still remains: your concept and my concept, understood as psychological contents, would not be the same. They may not even be similar.

Some philosophers and psychologists are impatient with air-fairy inner objects like concepts and images. What matters, what is real, is behaviour. So they argue that the meaning of a word is the influence it has on behaviour: words, like natural signs (such as clouds), bring about certain effects in the hearer. It is these effects that are the meaning of the words. Alas, this behaviourist account creates more problems than it solves. For example, words – even straightforward ones such as 'dog', never mind less straightforward ones such as 'the' – do not always bring about behaviour or even predisposition to behaviour; and similar behaviours may be brought about by events that are not words. What is more, the behaviour theory overlooks the difference between linguistic and natural signs. Clouds may make one believe it is about to rain, and the utterance of the word 'rain' may do something similar. In the former case, the sign merely happens, and happens to have meaning to someone; in the latter case, the sign is manufactured in order to have meaning. While natural signs mean, utterances have *meant meaning*: the latter can be received only if I see *that* it is meant, that *this* is what is meant, on the basis of the producer's apparent intention. And of course one can use the word 'rain', as I do now, without it altering everyone's expectation of getting wet or causing them to reach for an umbrella. My use is mere mention.

This kind of observation compels us to recognize that it may be futile to look for the meaning of words in isolation. They make sense only in the context of utterances – propositions, sentences, assertions, originating from a particular person at a particular time and place. On this basis, the meaning of words becomes *that which would have to be the case for the words to be true* – the so-called 'truth conditions' of the assertion they are used to make. The problems with this theory are manifold: it has abandoned trying to determine the contributions of individual words; it does not deal with the 'intendedness' of meanings; and it threatens to deflate both the concepts of truth and meaning in a few platitudes.[4] What is more, it does not deal with those perfectly meaningful sentences that are not statements of fact, to which it is not appropriate to apply the concepts of truth or falsehood – for example, questions and commands and utterances that are expressive of emotions.

The search for the meaning of meaning seems to be heading into a thicket of confusion. One way of dealing with this is to embrace the thicket. Under the influence of Wittgenstein, many philosophers argued that the meaning of a word or a sentence is its *use*. This seems to get round the difficulty that words have different kinds of uses – the meanings of 'Fido', 'dog' and 'the' are clearly different in kind. And it also copes with the fact that utterances may have many different purposes: words clearly have different uses when one is informing someone as opposed to telling them to do something, or swearing at them. Swearing may or may not be justified, but it cannot be true or false. While this theory avoids the problem presented by the wonderful variousness of words and in which they are employed, it is somewhat vague. Of course the meaning of the word has a lot to do with its use, but what does 'use' mean? And how does a word come to have this use unless it has a certain meaning or range of possible meanings? Isn't it because 'the' has a different meaning from 'Fido' or 'dog' that it can be used differently? We seem to have fallen into a circular account of meaning: the meaning of the word is the use to which it may be put, but the use to which it is put will depend upon its meaning.

The literature around the use theory of meaning expanded and became ever-more technical as the meaning of meaning receded before a growing army of philosophical drones. One of the discoveries that took philosophers (but no one else) by surprise was that words were used to *perform acts*: utterances were *speech acts*. The brilliant J. L. Austin pointed out that we utter words with a certain meaning in order to do something – for example, to warn, to inform. This *illocutionary act* is in turn the basis of something that we achieve *by* doing something: the so-called *perlocutionary act* that persuades, enlightens, upsets, frightens off, etc. The most developed speech-act-based theory of meaning was William Alston's notion that the meaning of a statement is its *illocutionary act potential*: two statements have identical meanings if they have identical illocutionary act potential.

This does seem to be an advance, but, as many philosophers have pointed out, the notion of 'illocutionary act potential' is not entirely clear. Alston meant the term to capture the contribution a word might make to shaping the illocutionary force of an utterance, but it remains tantalizingly vague. What will by now be clear, however, will be why the theory of the meaning of words, at least as pursued within the philosophy of language, which obsessed much of the professional philosophical community, as well as this amateur, for so long has lost much of its promise and a good deal of its rather specialized charm since the middle of the last century, which was when these accounts of meaning were jostling for position. The mystery of language seems to thicken the harder you look at it, perhaps because we have to use language to try to illuminate it. There has been an enormous amount of work on individual theories, but little of enduring interest has resulted. The reader may therefore wonder why we have entered this digression into analytical philosophy. It is because how words mean things – which drew me into thinking about pointing – remains one of the most fascinating questions about human consciousness. Pointing seemed to offer one way of approaching, even modelling, the link between words and things. Let us look at the most obvious link between pointing and the meaning of meaning.

Ostensive Definition and its Limitations

The connection between the big question – the meaning of meaning – and the small gesture – pointing – is pointed up in one pretty well obsolete word: *ostension*. 'Ostension' is worthy of attention because the wrong trees barked up by myself and others with its assistance are fascinating. And, what is more, when it comes to something as slippery and elusive as language – particularly slippery and particularly elusive because we have to grasp it using language – we may need to proceed by indirection. Getting language wrong, and seeing how one has gone wrong, may be a way of seeing a bit more clearly what it is, even though completely getting our language-buzzing head around it may escape us.

The problem of the charmed circle of language arises in two

contexts: when we think about how a child manages to acquire language primarily from its parents, without already understanding language in order to know what its parents are talking about; and how we can ever get to the meanings of words in a second language we are trying to learn without simply following a large trail of words that eventually circles back on itself. Using words to define the meanings of words to a child may be signally unhelpful, given that the words used in the definition are often more complicated, abstract or sophisticated than the word we are trying to define. Indeed, the more basic the word, the greater the discrepancy between the definition and that which is defined. If, for example, one looks up 'cat' in the *OED*, it is defined as 'a well-known carnivorous quadruped' – well-known, certainly, but rarely as a 'carnivorous quadruped'. In his recent account of trying to learn the language of an isolated Amazonian tribe, Daniel Everett identified a vicious circle: 'In order to learn their language, I must learn their language'.[5]

Ostension seems to be one way of breaking into this closed circle, in that we can, seemingly, directly indicate the meaning of the word without resorting to other words. 'Ostension', according to the *OED*, is 'the act of showing, exhibition, display, manifestation'. The noun is derived from a rare verb, to 'ostend', meaning 'to show, reveal, manifest, exhibit'. So-called 'ostensive definitions' were christened by the Cambridge logician W. E. Johnson, a brilliant teacher who introduced many new concepts to philosophy but who was reluctant to publish his ideas.[6] A non-ostensive definition of ostensive definition is 'Definition by an example in which the referent is specified by pointing or showing in some way'.[7] As the definition implies, there is more than one way of performing an ostensive definition. I might, for example, define 'pinching' by pinching a person's skin.[8] The canonical form of ostensive definition, however, is by pointing to a particular object, which is why we are discussing it now.

What we have already discovered about the nature of meaning should prevent us over-estimating what might be achieved through pointing-mediated ostensive definitions. They seem unlikely to work on their own even in the case of those words that appear to be amenable to a 'Fido'-Fido account of their meaning. Just how little unassisted pointing might achieve is illustrated by its insufficiency even in the case of proper names such as 'Fido' that should be tailor-made for ostensive definition. As John Searle observed, 'At first sight nothing seems easier to understand in the philosophy of language than our use of proper names: here is the name, there is the object. The name stands for the object.'[9]

But as Searle further observes, things are not at all simple: proper names are not the pure referrers they seem to be, because we can 'never get referring completely isolated from predication'. To do so would be to transgress those principles by which we target language on particulars.[10] Although stand-alone pointing is proto-referential, it does not enable us to achieve full-blown reference by the back door, so to speak, nor lay the groundwork for future reference, because stand-alone pointing cannot replicate those abstractions – the predicates, the universals, the general terms – through which reference is usually mediated. The pretty story that we can define expressions ostensively by pointing directly to their bearers, because carnal pointing replicates verbal pointing, does not hold up, therefore, even for the most seemingly straightforward cases.[11] To put it another way, if the meaning of 'Fido' *were* something that was hairy and barked, it would be very difficult to get it to combine with other meanings in a sentence such as 'I am fed up with Fido'.

And yet there is still something overwhelmingly attractive about the notion that meanings of words are something you can point to, and that ostension will enable you to pin language to the physical world; that the index finger can cross the gap between the things that are around us and the realm of discourse that creates that vast rumour called the human world. Just how attractive the notion is may be demonstrated by the quality of the minds it has seduced. Take the late Lord Quinton, a man of immense and subtle intellect who had grown up intellectually in the unforgiving environment of Oxford philosophy in the middle of the last century, a world populated by some of the sharpest and most competitive minds.

In a paper published over forty years ago, Quinton contrasted indirect with direct definition.[12] Terms, he wrote, may be given meaning indirectly 'by definition in terms of other expressions, already understood' (that's the dictionary approach, whose limitations we have already seen), or they may be defined directly by ostension. This is the view we are examining and it is not original to him. What is new is his account of what ostension does, an account which digs deeper and reveals what is at the heart of faith in the power of ostensive definitions and, by default, certain properties of verbal communication. According to Quinton, ostension correlates words with 'observable features of the world'. Nothing too revolutionary there. He then goes on to say that ostension is the final guarantor of meaning for all terms except certain terms in logic, which are defined by other terms in logic. A non-logical term is one 'whose meaning is *wholly* determined, in the end, by ostension'. And what precisely does ostension do? It 'correlates [expressions] with particular regions or features of the extra-linguistic world', and, more specifically, and more relevant to pointing, it 'correlates a term with a class of spatio-temporal regions'. And that is the punch line: ostension connects language with the world it expresses by correlating words (he calls them 'terms') with types of bits of the material world.

Could one really do this by pointing? Would the index finger really be able to co-index the material and human world, the world of space-occupying objects and bits of the semantic field indicating non-localized general possibilities? Once we put it this way, Quinton's assumptions look highly vulnerable. Yes, we can point to individual objects and, yes, objects occupy certain spatial regions over a certain period of time. By pointing to individual objects, I clearly point to the space they occupy. I don't, however, point to the temporal region they occupy. The pointee is always the object insofar as it is present and *in the present*. The tense of pointing is *now*. I cannot point to the future or the past of an object. At best, therefore, I point to a time-slice of the object that corresponds with the period it is being pointed at, the time it spends as a pointee.

Even if we modify Quinton's account to say that ostension correlates a term with a class of *spatial* regions, this would still require of pointing more than it could give. For pointing to an object does not of itself give us a *class* of objects. First, it is an additional step to make the object a member of a class, an entity instantiating a class. Secondly, any given object may instantiate any number of classes. There are two reasons for this: first, we may access a given object by many different expressions, as when I refer to a cat as 'cat', 'pussy', 'Felix', 'that bloody animal', etc.; and second, a given object may be used in the ostensive definition of any number of expressions or terms. If, for example, I point to a cat, I might use it to instantiate 'cat', 'animal', 'pet', 'furry creature', 'fur', 'black' and so on. This systematic ambiguity, which has been noted by many philosophers, most famously Wittgenstein, is not an accidental feature of language. Words get hold of objects by *aspects* of them and any given object will have many different aspects.[13] A given term will express one aspect of, or one 'take' on, the object. The finger, however, can point only to objects as a whole: it has perforce therefore to point to all of the takes on the object. More precisely, it cannot discriminate between takes.

All that pointing does, after all, is to make explicit a particular direction radiating out from the producer of the sign. Deciding which, of several candidate objects lying along the direction line, is the pointee, or appointee, requires further information. In the absence of speech, the pointee needs to be self-highlighting, or at any rate, some assistance is required from the object if we are to identify it as the pointee. This might be forthcoming if, for example, it is moving. A succession of pointings towards a moving cat, or a pointing that is locked to its trajectory, will make clear what it is that is being pointed to. We may express this by seeing the successive, or sustained, pointings as a class of indications that have a common object and by this means enable the consumer to pick out that object.

Such a mode of highlighting is clearly not available when what is pointed out is a *class* of objects or, to use Anthony Quinton's phrase, of regions of space or space–time. The movement of the cat across the room may highlight the pet, but not the class to which it belongs. And expressions, with the exception of proper names, are of a general nature. If they refer to individual objects, they do so through the class which the object is seen to indicate. The descent from generality to singularity is achieved through restricting the range of possible candidate members of the class. These means will be ultimately extra-linguistic, though they may be mediated by more language. Most directly, we may physically restrict the universe of discourse: the range of candidate objects can be confined to the here and now of the speaker, to her sensory field. Under such circumstances, pointing may further narrow the range of candidates by a drastic reduction of the portion of the sensory globe that may field candidates.

There is something deliciously odd, even perverse, about ostensive definition. When I point to a cat to give you the meaning of the word 'cat', I seem to access meaning through reference, or proto-reference. This is, of course, the wrong way round. Normally, the meaning of a word is that by which we access what it is referring to. You say the word 'cat' and I, understanding the (general) meaning of the word, look for a particular instance. The particular instance is the referent of the word on this occasion. When you point to a particular cat to give me the meaning of the word 'cat', meaning is accessed through proto-reference. Of course, referents such as individual cats are particulars and particulars both fall short of general meanings (such as belong to common nouns such as 'cat') and exceed them, because they have an indefinite number of properties not specified in the meaning of 'cat'.

This reversal of the relationship between meaning and reference, accessing meaning through reference achieved through pointing, should put us on our guard and lead us to expect what we have found: that ostensive definition acting on its own fails to capture a genuine referent even in the easiest of all cases, proper names, where the meaning seems (misleadingly) to be identical with a referent and the latter to be located in a portion of space that fingers seem apt to point to or out. If it is inadequate even for those words that denote things that separate themselves and stand up to be pointed to, it will hardly do for words that designate classes of objects, or abstract objects, or general properties.[14] And it will be even less helpful for function words (conjunctions such as 'and', and articles such as 'the') and terms such as adverbs and prepositions that manifestly do communicative work only in conjunction with other words in the context of an entire sentence or statement. The truth is, all we can achieve through pointing is to identify which, out of a crowd of (verbally) pre-specified entities, is the entity in question, as when, pointing to a group of people, we indicate which one is 'Fred'. Most of the work of ostensive definition has to be done beforehand; other non-digital indications are necessary to shape the attention that pointing targets. Pointing is at best the final stone on a pyramid of prior linguistic preparation. It is only because this prior preparation, the world-picture and the particular fix within it, is 'off-stage', that we overlook it and consequently exaggerate what can be achieved through pointing.

And yet the relationship between pointing and language remains elusive and intriguing. It is difficult to shake off the idea that this relationship promises to tell us about the deepest and most mysterious of all relationships, that between verbal expression and the material, or at least extra-linguistic, world. Pointing is the most blatant example of 'deixis', a property that connects signs with the material circumstances in which they occur. It seems to be a place of intersection between carnal and, consequently physical, space and the abstract space of discourse. Pointing silently utters 'That...that thing...that state of affairs' and links that local explicitness with the massively elaborated explicitness that is made possible through language. This intersection also reminds us of the extraordinary fact that abstraction – which lifts us above our condition – begins with the organism and is carved in exhaled air. Ostension makes this explicitness itself explicit. No wonder it is almost impossible to leave it alone, and no wonder philosophers, from Cratylus to St Augustine to Wittgenstein, have found it fascinating. Indeed, Wittgenstein's engagement with St Augustine and with the latter's over-estimate of what can be achieved through pointing was a decisive event of twentieth-century philosophy. It occupies the opening passages of Wittgenstein's *Philosophical Investigations*. (Wittgenstein admired St Augustine enormously for his tormented seriousness and his intellectual passion, seeing in him a fellow spirit, whom he could admire at the tolerable distance of 1,500 years.)

Feeding the Word-Hungry Infant (1)

Humans are unique not only in pointing – pointing at and pointing out – but in teaching their young. Other animals may learn from their elders but they do so by imitation; their parents do not deliberately manufacture examples of behaviour in order to instruct their offspring how and why to perform certain actions. Even less do animals teach by precept, in a space of shared consciousness. We humans, by contrast, teach our children to speak, or, more precisely, we actively assist at an ongoing process of language acquisition. It is therefore not surprising that the idea that language can be, and indeed is, taught by means of pointing, is inescapable. And this is where Wittgenstein engages with St Augustine.

Wittgenstein begins his *Philosophical Investigations* with a quote from St Augustine's story of how he as an infant learned to speak: 'When they (my elders) named some object, and accordingly moved towards something, I saw this and grasped that the thing was called by the sound they uttered when they meant to point it out.' This theory of language acquisition, Wittgenstein observes, gives 'a particular picture of the essence of language': 'individual words in language name objects – sentences are combinations of such names. – In this picture of language we find the roots of the following idea: Every word has a meaning. This meaning is correlated with a word. It is the object for which the word stands.'[15] This picture has one element which we have already demolished: the '"Fido"-Fido' notion that words mean objects and the objects they mean are the meanings of those words. It also has other claims from which we have also dissented. Most importantly, among these is an erroneous 'atomic' notion of language (corresponding to an atomic notion of the world) in which isolated words have stand-alone meaning. It overlooks how words belong to a system. Language is not merely what that early neuro-biologist of language, the Victorian neurologist John Hughlings Jackson, called a 'word heap', in mockery of those who treated it as if were a collection of terms.

Fortunately for St Augustine, his reputation does not depend on his theory of language-acquisition, given that it fails to take account of so many things that we have already noted: the interconnectedness of language, such that individual words have their charge of meaning only in relation to other words, and words deliver only as part of larger units such as uttered sentences; the multiple use of words in language, of which denotation is only one; the varied uses to which language is put (swearing as well as explaining, teasing as well as describing); and so on. Most importantly, he overlooks what, as we have seen, is necessary to make an ostensive definition provide the meaning or function of a word. To 'get' an ostensive definition, many other things have already to be in place: ostension can act only as a coping-stone completing an arch. More specifically, the Augustinian model suggests that acquiring an understanding of language is rather like switching on a series of separate light bulbs, whereas in reality it is more like what Wittgenstein later spoke of as 'light dawning gradually over the whole'.[16] Even the placing of coping-stones is an iterative process: the parent will expose the child to repeated ostensive definitions of the same word; the child will use the word experimentally and sometimes get it wrong; and, what is more, it may actively lead the process.

Even so, pointing does have some role in language acquisition. Many parents will recall with mixed emotions long car journeys in which early speech learners will point to object after object and say 'Wazzat?' The toddler will have grasped the notion of something being *called* something, a necessary precursor to learning the particular name of an object, or grasping fully that objects have names. Under such circumstances, the child brings half of the ostensive definition, by picking out the object. The parent then has to provide the naming half. This can often be a frustrating experience for both parties, as the burden of identifying what the child is picking out lies with the parent. Answer after answer to 'Wazzat?' can prove unsatisfactory because the parent has failed to get the point – or the pointee. (This does illustrate how the journey between word and object may be two-way: under certain circumstances, the object is picked out in order to give the meaning of the word; in others the word is uttered in order to label the object.)

The limitations of ostensive definition are even more apparent when we appreciate how quickly things get complicated. Even very young children capable only of ejaculatory, single-symbol utterances can use words in a non-situation-specific way: 'The word *kitty* may be uttered by a baby to draw attention to a cat, to inquire about the whereabouts of the cat, to remark that something resembles a cat, and so forth.'[17] There is no way that ostension could capture this multiple use of individual words. It is perfectly obvious that all words are caught up in a network of complex communicative skills, built up over countless iterated interactions that gradually awaken in the infant.[18]

Ostensive Definition and its Limitations (2)

The limitations of ostensive definition are interesting because they highlight the troubled relationship between what may be called the universe of discourse (and its most exclusive *arrondissement*, the space of reasons) and the space of the natural, or at least physical, world that surrounds us. The problem of how the two connect is infinitely fascinating. There is something deeply, and fatally, attractive about the notion – to which I once subscribed – that ostension can achieve this, as if one can skewer a bit of the semantic field on the end of a (material but meaning-bearing) digit pointing to an object in the material world, so that fluttering meaning and stable object are fastened together as if by a pin. This is clearly wrong, but it is still difficult to relinquish the notion that some kind of direct connection between words and the physical world is necessary to ensure that the language remains, as it were, on the gold standard so that utterances stay in contact with extra-linguistic reality. Without such contact, there seems to be a danger that words will drift off into a world of their own and we might even have to accept the kind of nonsense purveyed by the structuralists (and their post-structuralist and other postmodern successors) that words are not 'about' anything other than themselves and that there is (to echo

Jacques Derrida's famous formula) 'nothing outside of the text'.[19] In short, the danger of denying a direct connection between language and extra-linguistic reality is that we might succumb to a form of linguistic idealism, an extreme and unsustainable form of the anti-realism that has none the less managed to live a quiet respectable life in some quarters of Anglo-Saxon analytical philosophy. Ostensive definition by pointing seems to provide a no-nonsense link between the words we speak and the objects we bump into.

As we have seen, one of the intuitions that leads philosophers to over-estimate the role of pointing in bridging the gap from babbling to speech is that there is a kind of one-to-one relationship between words and things, so that pointing can pinpoint the meaning of words by picking things out. Although we have identified numerous reasons why this is wrong, it remains a fascinating idea, particularly when it is developed in the way that Wittgenstein did in his early 'Picture Theory of Propositions' in the oracular *Tractatus Logico-Philosophicus*, that thought it had the last word on the relationship between language and the world and by this means was able to bring philosophy itself to a conclusion.

Wittgenstein argued that there was a one-to-one relationship between the components of a proposition such as 'The cat is on the mat' and the components of the states of affairs it expressed. The proposition succeeded in expressing the state of affairs in virtue of having a *form* in common with it. More precisely, the proposition and the state of affairs have elements that can be mapped on to one another because they have the same 'logical form'. The proposition worked ostensively: it *showed* its sense. This theory has been entirely discredited, not least by Wittgenstein himself, for reasons that will by now be apparent. It doesn't even work for simple declarative statements uttered without ulterior motive – such as 'The cat is on the mat'. The use of the word 'the' makes that clear enough. The enduring value of the theory, however, like many of the wrong ideas of great thinkers, lies in its exposing, by taking to their logical conclusion, something that we feel about the nature of the relationship between language and that which is expressed through it. The 'Picture Theory of Propositions' uncovers what it is that underpins the over-estimation of what can be achieved through ostensive definition.

Although it is clear that pointing is a succession of disconnected symbolic gestures, while words always have to be connected with one another (to form sentences or phrases) in assertions and other speech acts, and that individual words have meaning only as part of a *system* of signs, it is almost impossible to escape the feeling that pointing is a paradigm of how we might map linguistic coordinates on to physical ones; to provide places at which, at the very least, pointing tacks the *verso* of discourse on to the *recto* of the physical world.[20] In reality, the gap, ultimately, between material objects and language is between the denizens of the universe of indexical awareness – physical objects – and those of the universe of deindexicalized awareness. The inhabitants of the latter exist as the possible realization of generalities: they are not located with respect to the objects in the material world, least of all the body of the speaker. Consequently, uttering the word while pointing to an object cannot make that object an exemplar of meaning, gathered into the universe of deindexicalized awareness while at the same time guaranteeing that it will retain the singularity it has as a piece of matter. This failure underlines how our human world has two profoundly different faces: that of a collection of material objects and that of an ocean of collectivized awareness gathered up in language. Ostension, reaching into the world of material objects, cannot directly touch or determine a location in the deindexicalized realm housed in language. Something more than pointing is required to connect one realm with the other. Putting an object in italics will not translate it from the material to the semantic realm.

Feeding the Word-Hungry Infant (2)

There is a limit, therefore, to what pointing, or, rather, ostensive definition using pointing, can achieve. We must not, however, dismiss it entirely. The notion that preverbal and verbal life can be linked to some extent through preverbal quasi-reference through pointing has both intuitive attraction and empirical evidence in its support. The declarative act of pointing, where the infant draws the adult's attention to something the latter may not have been aware of, seems to overlap with a central function of language: to take hold of something that is absent, via a sign. And the fact that the act of pointing is not an event that occurs spontaneously, like a twitch, and is then appropriated for a secondary, communicative purpose, but one that takes place only to perform an act of communication means that it is on the far side of the line that separates natural signs that occur independently of communicative meaning (as when clouds mean rain) from verbal signs that exist only to mean – that are *meant meaning*.

There is also something referential about pointing and pointing to something invisible to another awakens in that other person's mind the sense of a general possibility, which is the very essence of what is conveyed through language. There is empirical support for the notion of pointing as a transitional form of behaviour linking the babbling infant with the talking toddler. As a corrective, it may be important to glance at some of this empirical data.

Pointing closely precedes language, and studies have shown that the amount of pointing at twelve months predicts speech-production rates at twenty-four months. What is more, the earlier the onset of pointing, the greater the amount of speech at fourteen to fifteen months.[21] There is a deeper empirical connection, a link between the onset of pointing and grasping the not at all obvious idea that objects have names; indeed the two seem to occur in the same week. Very soon after babies have acquired the art of pointing, they use it to individuate an object within categorical perception. And soon pointing and language are working in harness: combinations of pointing and a word – 'Doggy!' – are seen consistently before sixteen months, just before the child begins to make two-word utterances: 'Mummy, sock!'

As George Butterworth says, 'pointing serves not only to individuate the object, but also to authorize the link between the object and speech from the baby's perspective'. And in this respect it is important to note that parents more often label objects children point to than they do ones the children reach for: they turn pointing into a request for the name (the general sense) of the object, not a request for the object itself. This is underlined in a recent study that St Augustine would have approved of, published in the hugely prestigious scientific journal *Science*. The authors built on their earlier work in which they found that when mothers 'translated' their child's pointing gestures into words, those words tended to become part of the child's spoken vocabulary several months later. They found that the parents of children from higher socio-economic-status families used more pointing gestures, as did their children, than children from lower socio-economic-status families. Intriguingly, even when differences in speech interactions between parents and children were controlled for, this difference correlated with the size of the child's vocabulary at school entry age.[22] As is so often the case, pointing reaches into the far-distant future. All those gruellingly tedious 'Wazzats?' are worth engaging with. At any rate, for the child, pointing brings cognitive rewards, as well as the tangible ones that reaching reaches for: an abstract sweetmeat that begins, but does not remain, in the mouth. Establishing the link between object and speech, recapitulating Adam's primordial linguistic act of placing a grid of words over the world, is a complex iterative process, with pointing assisting the understanding of verbal reference and verbal reference clarifying what is being pointed at. And of course pointing is not rendered superfluous by language, laid aside as a prop of infancy. It is integrated into verbal, and other nonverbal, signs, as the most powerful and versatile of paralinguistic tools. It is, as linguists describe it, an 'interstitial' practice – a bit like the 'pointing' that keeps the bricks together in a house.

So, while St Augustine overstated the role of stand-alone pointing in language acquisition, ostension does allow visible objects to be associated with certain word sounds and 'this is the royal road' (but not the only route, and it is metalled by other cognitive activities and capacities) 'to language'.[23] While there is no way that the complex syntactical and semantic system that is language could be pointed out, even less connected to the world with deictic pins or tacks fired by the index finger (these need a general *nous* that enables an intuition of rules of sounding and meaning and a long process of experimentation, correction, play, and communication), pointing does play an important role in bringing language to bear on the here and now. This is essential if words are to open up the sense of the there and then, and the general elsewhere and the no-when, that is constitutive of the world that humans have built in parallel to the natural world. While it is too much to expect of the unaided index finger, even one animated by the complex intentions of a human being, to point literally to the link between the words and the meanings they have, it is clearly an enormously important aid to the process of inducting a child into the common human cosmos, to the awakening out of the mere sentience of the organism to that place of sense where we find much of our happiness and sadness and all of our collective power to act for good or ill.

chapter six
Pointing and Power

The Power of Pointing

The index finger indicates by means other than pointing. It can be used to trace round things to mark their shape. It can do this in subtle, indirect ways, using vague symbolic gestures. The cruel way of signifying that someone is mentally disturbed by twirling the index finger next to the forehead gives bodily expression to the abstract idea of someone whose mind is supposedly spinning. We may use the forefinger to assist quantitative tracing, as when we enumerate the items around us, relying on the digit to tell us where we have reached in our counting. The forefinger may help us keep our place in the most abstract reaches of the deindexicalized realm, as when we move down a column of fingers or pinpoint a place in an index, as if we were co-indexing the abstract world of themes and the physical space in which the terms that refer to them are set out. Those who have difficulty following written texts will run their fingers along the lines of print, as if keeping their attention fastened to a target they themselves are moving.

Pointing, however, is the big one and we are far from exhausting its many facets. In concentrating as we have done on declarative pointing, we have not done justice to the gestural versatility of the forefinger and the richness of its indications. We shall explore this in the present chapter in which we have to acknowledge that while pointing is beautiful as well as useful, it can also be very ugly and very irritating.

Just how ugly and irritating was highlighted for me by a triumphant Alastair Campbell, Tony Blair's press-secretary, when he was interviewed by an obsequious BBC after the publication of the report of the Hutton Inquiry into the suicide of the scientist David Kelly. The report, by a breathtaking bouleversement, left in the clear a government that had launched a war on the basis of what many regarded as a foundation of misinformation and untruths, a war that has so far directly or indirectly caused the deaths of approximately 600,000 Iraqis. The BBC, which had merely drawn attention to this shaky foundation, was hit by a tsunami of abuse. Mr Campbell was at the heart of the government, and at the heart of selling the case for this disastrous war. He responded to his interviewer fluently, indeed passionately, rejoicing in his vindication. By way of emphasis, he raised, he wagged, he waved, he wiggled, his index finger. As I watched him, it seemed to me that this single digit stood for all the arrogant, opinionated, moralizing, morally impervious people I have come across in my life.

Finger-wagging puts what is being said in italics; it raises the font of warnings, commands, menaces, criticism. It permits even the softly spoken to shout that you are naughty, stupid, wicked, silly. The wagged finger asserts authority. It says: 'You mark my words' and mark it you will. It says, 'I know, you don't know' and 'Woe betide…' (and it is only under such circumstances that in modern English we 'betide' anything). The raised finger says, 'I will say this only once, so listen up.'

Those who like making obvious, but dubious, connections link the erected index finger with the phallus, the root of power, dominance, threat and oppression in a patriarchal, 'phallocentric' society. Those less easily seduced by symbolic associations will note that the index finger is as often raised by women as by men. At any rate, it serves both sexes to command – attention, authority, and control of meanings. Its powers are astonishing and can be exercised so easily. You are chairing a meeting and X on your left is talking almost uncontrollably. He has something, everything, to say about every item that comes up. On your right, Y is sitting there patiently. He raises his index finger silently and keeps it raised. His standing request for attention, and his evident patience, eventually give you the strength, and confidence in your own authority, to silence X and ask him to allow someone else a bit of airtime. While the little gesture of raising one's index finger may be a way of requesting permission to speak by signalling that one wishes to say something, it has more active power if one is, say, chairing the meeting rather than merely attending it. The raised finger is somewhere between requested and demanded silence. It can be almost as annoying as finger-wagging, as I discovered when I was on a committee chaired by a completely incompetent individual who wanted to talk all the time. That she did so with her eyes closed – to indicate the intensity of her concentration – was less annoying than the fact that she had her right index finger permanently raised as she spoke, as if to pre-empt interruption.

Part of the power of the finger is that it both asserts and effaces itself at the same time: it has an eloquent silence; it says and does not say 'I'. Through its assertive silence, and the fact that it looks like the digit '1' and is single, it occupies a space rather like that of the personal–impersonal 'one'. Writing of 'one', *Fowler's Modern English Usage* observes that using 'one' permits the self-conscious writer 'to be impersonal and personal at once…here is what he has longed for, the cloak of generality that will make egotism respectable'.[1]

Finger-wagging and air-jabbing are some of the ways that speakers demand the attention they feel they deserve and may not otherwise command. Thus does the index finger signify the sickening mystery whereby some people feel the right to boss over others. It is self-legitimating and, raised, it locates the owner at the centre of the conversational field, the pole round which the other speakers dance. Finger-raisers will often be deeply offended by the others who interrupt their flow.

My allergy to this kind of behaviour (as I suspect may be showing more than a little) is so strong that it extends to relatively innocent targets. For example, one of my favourite mediaeval philosophers, Duns Scotus – the inspiration for, among other things, the verse of Gerard Manley Hopkins, one of my favourite poets – is often portrayed holding his index finger to his lips. He appears to be 'enjoining' (the word tends to be used in this context) silence. I try to soothe my irritation by thinking of the extraordinary nature of this gesture: the command not to emit words from one's lips issued silently, by touching the place whence words arise. Alas, this cunning way by which speech can be circumvented in order non-contradictorily to commend silence, does not palliate my loathing of this gesture. It is an arrested finger wag, negating in advance all that the listener is about to say, and in retrospect all that he has said so far.

Stabbing the air, even if the index finger is erect, is a symbolic assault on one's person: a mode of exasperated shouting that prods the interlocutor in order to expropriate her attention. It force-feeds the speaker's version of events onto the other person, who is thought to be resistant or dim-witted. It may be the harbinger of someone losing control. Such pointing is poking at a distance, a one-fingered punching, aimed not at the body but at the self and its esteem, and it may prefigure actual poking that reduces the person to his despised body, denying the inwardness by which the self can escape from the defining, diminishing, finger of the other. Air-stabbers may be close to poking you in the chest, poison-tipping their gesture with your pain or discomfort at unchosen intimacy. For pointing can be worse than egocentric passive-aggression: it can be aggression pure and simple. The pointing finger can be used to nail laughter, or menaces, to their targets. The victim feels skewered, as if by an arrow or a digital gun.

The efficiency of pointing at reaching its inner target is indicated by the fact that the symbols in sign language for 'I' and 'thou' are pointing gestures. And that we indicate ourselves by pointing to our bodies. When you point to me, you do not simply indicate my body: you touch my person. Aggressive pointing is a digital glare; but it is more than that. The poke in one's consciousness threatens to become something more hands, or fingers, on. It points to direct contact, the poking and prodding that makes life for some such misery in institutions: the psychological assault that passes into a physical one. Under such circumstances, one is nailed to one's non-transcendent particularity, becoming whatever it is that is under the finger of one's assailant. The extended index finger may pack more punch than the closed fist.

To be pointed at is to be picked out, to be picked up, to be made available: 'Hoy! You!' Pointing-as-summons can take the form of beckoning, in which the index finger strokes the intervening space or milks you into compliance. The teacher beckons you to the front of class. He need not say your name, merely gaze at you. That crooked finger of his makes explicit the power that has you in its grasp, is an active bar of the cage that is the school whose rules you must obey. The slower the beckoning movement, the more sinister it feels. The beckoner can be patient because he knows that sooner or later you will obey. It is a kind of one-finger grasping. Because his authority is irresistible – he is bigger, more senior, more powerful, more terrifying – one finger is enough.

To be beckoned is to be hooked and drawn away from the world in which you are the centre to one in which you are a mere unit. You are *hailed* and, as the tragic, insufferable Louis Althusser (one of a select group of philosophers who have murdered their wives) wrote, such 'interpellation' makes you a 'subjected subject', even an 'abject'.[2] Isolated pointing is not, of course, enough. At any rate, it cannot act on its own: the space into which we point, constructed out of the infinity of pointings, points back to us. You are 'always already' (to use a favourite phrase of the post-this-that-and-the-other philosophers) pointed out; but some pointings may indeed be life-changing. That famous poster in which Lord Kitchener, spokesperson for King, Country and Empire, to which you owe everything, pointed you out and informed you that 'Your Country Needs You', summoned you to give everything to the collective cause. Calls under such circumstances are clarion because they are underpinned by the assumptions which define the world in which you live. They are not easy to resist. Your sense of duty and your sense of what you are converge fatally in a response that will lead you to the misery of life in the trenches with or without death. On such occasions, the pointing finger, which nails what we are individually to what we are collectively, shows us that the society in which we live cannot be rejected without loss of our sense of ourselves; that there is no outside that is not also inside – inside the framework of the assumptions that we require to make sense of ourselves, and the world. (Which is why, under certain circumstances, not to be picked out can be equally unpleasant, as some of us will remember from school sports.)

No wonder that, when we are pointed at, we feel both indefensible and the need to defend ourselves – in the position of one who *s'excuse* and hence *s'accuse*.[3] Not for nothing is 'pointing the finger' the prime metaphor for accusing someone of a crime or misdemeanour – even the crime of being different, of sticking out like a sore thumb. Full marks then to our parents, who taught us that it is rude to point.

Why it's So Rude to Point

It is easy to see why some modes of pointing are rude but less easy perhaps to grasp why pointing can sometimes penetrate us so profoundly. Why has this chapter been so hysterical? Why is it *so* rude to point at someone, even if the action is not meant to be cruel or demeaning, is not accompanied by laughter, even when the pointing finger is not guiding jeers to their target, allocating blame, picking us out of a reluctant crowd for some unpleasant, dangerous, or humiliating task? It is because the pointing finger prods at a vulnerability we all share. We are skewered on the attention of another person and any others to whom the pointing is also addressed. T. S. Eliot's

J. Alfred Prufrock, whose overwhelming sense of inadequacy makes his love-song so poignant, knew how one is 'pinned and wriggling on the wall'. We are fixed 'in a formulated phrase', the object of another's gaze and, for this reason, are subordinated to their self-consciousness.[4] We acknowledge them but they acknowledge us only on their terms, reducing us to a part of ourselves which is both much less than what we are, and represents us in a poor light. Pointing not only exposes one to an especially visible gaze, it is a gesture that invites others to gaze and so to co-opt their self-consciousness to a reductive account of one's own self: the woman who fell over, the man with the funny look in his eye, the boy caught with his trousers down. Pointing, in virtue of co-opting other conscious-nesses, intensifies the sense we all have at times of being known and yet not-known – of being 'mis-known', of helpless exposure to uncomprehending eyes that imagine they comprehend us.

Being comprehended by what is only a small part of one's life is an exquisite form of social torture. If knowledge is power, to feel that one is known is powerlessness. The pointing finger, by a miracle of malice, not only prehends, in the sense of grasping and plucking, but also *com*prehends. The self enclosed in this way secretes another kind of carapace which offers no shelter, no mitigation of nakedness: a film of sweat in which one bathes in humiliation.

In being pointed out, one is the victim of turned tables. The invis-ible observer at the centre of the field, the *flâneur*, the pointer, is displaced from the centre and is himself located in the primordial away ground: another's field of observation, another's world. The spectator becomes part of the spectacle. This reversal of meta-physical fortune was something to which Sartre devoted many memorable pages.[5] The peeping Tom looking through a keyhole is master of the world that is laid out before his gaze, at the heart of which is the woman unconsciously offering the gift of her evolving nakedness. And then the boy himself becomes aware that he has been spotted. He, too, is an object in another's gaze, an object of the judgement of others. The king of his own sensory field is reduced to a naughty boy in the field of someone else's, or the world's, consciousness.

One of the most profound things ever said about human beings was an observation made by the nineteenth-century German philosopher G. W. F. Hegel in *The Phenomenology of the Spirit*. It lies at the heart of this early work, and its significance permeates the massive *oeuvre* he piled up over the next quarter-century. 'Self-consciousness,' he wrote, 'achieves its self-satisfaction only in another self-consciousness'; 'Self-consciousness exists in itself and for itself when and by the fact that it so exists for another; that is, it exists only in being acknowledged.'[6] Such self-consciousness, and the hunger that it creates for acknowledgement by other self-consciousnesses, distinguishes us from all other beasts. The most fundamental appetite of human beings, when they are not reduced by destitution to physiological need, is to be valued by another human being whom they themselves value.

At the bottom of this extraordinary appetite is the fact that we, alone among sentient creatures, have a sustained and pronounced sense that we *are*. Our existence is, as it were, in italics. I have already called this sense of our own being the existential intuition, which has made several appearances in this work, since it is a necessary pre-condition of our being the Pointing Animal. The existential intuition cannot, of course, be wrong. *That* I am must lie beyond the reach of any possible doubt, as it is I who must entertain, or be aware of such a doubt. As Descartes pointed out, even the very act of wondering whether one exists is proof that one does exist. It is sufficient to think or doubt that one exists to be assured or reassured that one does. Doubting one's own existence is an existential impossibility.[7]

It is a different matter entirely when one asks not *whether*, but *what*, one is. This is a tricky question for numerous reasons, but among the most obvious is that one is many things at any given time. The identity – the *what* – that one seeks to determine has to be upheld in the face of change. The quest for identity can never come to rest. *That* one is is inescapable; *what* one is is an unending quest, if only because the answering changes the questioner and the questioning influences the answer. And yet one wants to be *something* – a distinctive kind of thing that one can say that one is.[8]

There are reservoirs of stability to draw upon, to support the intuition that one is something at a given time and the same something over time. (The notion of identity has both of these components.) There is one's body, for a start, which changes only slowly and has a robust audit trail (spatio-temporal locations) linking its successive manifestations. This does not, however, quite answer the existential question. While the body underwrites one dimension of sameness – 'I am the same singular' – it does not underpin the other aspect of sameness – that I am of a certain stable character. It provides neither clue to the kind of self that I am nor, as must follow from this, assurance that I am the same kind of self over time. When we are talking about kinds of selves we are not referring to some natural kind. What kind of thing I am is at least in part constructed in the consciousness of another self, or the consciousness of the collective self, or, more precisely, the consciousness of another self who, at that time, stands for all the other selves, for a reference group against whom one measures one's self. The expectations and judgements of others are relatively stable – or they are relatively stable compared with the flux of impressions that constitutes the material of my unmediated self-consciousness. These expectations and judgements are in part deposited in, and super-stabilized in, the formal roles, offices, etc. that I hold. They enable me to respond to the questions 'What am I?' or 'What are you?' – with predicates: a doctor, Mrs T's husband, the father of her children, Chair of the Ethics Committee, and so on.

The appearance of stability, however, is just that: an appearance. And it doesn't deceive anyone, least of all ourselves. We look at its most impressive manifestation, the ten-page curriculum vitae, and realize how little this has to do with us, with how we are perceived, with how we feel about ourselves, with how good or bad we judge ourselves to be. The CV lists a multiplicity of scattered facts about us: date of birth, marital status, names and ages of children, qualifications, list of posts held, awards, publications, etc. Their multiplicity, and, worse, their almost comical heterogeneity, warns us that what we are reading about is nothing as unified as a self should be. Worse than this, even the individual facts seem to have little to do with us.

In his deeply moving novella, *A Boring Story*, Chekhov gives a first-person account of the life of an elderly eminent professor of medicine who is reaching the end of his career and probably his years.[9] His life is falling apart, but his reputation is undimmed. He is visiting a foreign city to give a lecture. He reads about his arrival, and his eminence, in the newspaper and says of himself, 'Here I am, a lonely old man in a hotel room, with an aching cheek,' and concludes that his name has acquired an independent life of its own.

This is an extreme example of how the things that assure us that we are definite, and continuing, the offices we carry, the things we achieve and which remain attached to our names, have a tendency to dry into carapaces that can easily be peeled away. What we feel we are and what it is that gives us our distinctive characteristics in the eyes of others are always separable, at least in part because, moment to moment, we do not quite fill out the words that attach to us and make up the relatively stable judgements we have, and need, and cultivate, in the eyes of others. And so we are vulnerable to being reduced by the index fingers of others, as they point us out. The bubble-pricking forefinger can existentially shrink us to our present moment, our present appearance, to some current action or activity, and this will always be less than we feel that we are.

Pointing and Signs that Point to Themselves

We have seen how powerful pointing is. The power of the index finger becomes even more impressive when we recall that it is the forefather of all those inscriptive devices – the reeds, sticks, quills and fountain pens – that have incised human consciousness on media more stable than the air from which we fashion our speech and the consciousness of our interlocutor. The way in which tools inspired by the primordial pointer have transformed our ability to collect together our individual understanding and thus immeasurably increased our ability to act on nature is an immense topic. Let me therefore pull back and simply reflect on the use of the finger itself as a writing instrument.

Peter Godwin, the journalist and author of several brilliant memoirs, when revisiting his native Zimbabwe where Robert Mugabe has reduced to the country to ruin, reports how, when he stopped at traffic lights, a child rushed out and wrote 'Help me!' on his windscreen in snot with his forefinger.[10] At the biblical Belshazzar's Feast, the writing on the wall foretelling the tyrant's fate was inscribed by an index finger. Less portentously, we all recall drawing figures and writing mottoes in the condensation on our bedroom windows on a cold winter's morning. And less portentous still is an example illustrating something that is implicit in pointing and has been just on the edge of our discussion of this wonderful gesture: its tendency to self-reference. The one who points points to herself. When I indicate something, silently saying, 'Look at this,' I am in fact saying, 'Look at this with me,' and hence 'Look at me!' Pointing points to a language that is riddled with meta-language.

This arresting example of finger-writing was once inscribed in the dust on the side of an unwashed lorry:

I WISH MY GIRL-FRIEND WAS AS FILTHY AS THIS

The 'this' referred literally, of course, to the filthy lorry and its dusty surface was the paper-and-ink that the finger pen could write on, or in. If we set aside the additional ingenuity that the ink in this case is negative – the letters are formed of the (comparatively) dust*less* track made in the dust – this is only to enable us to concentrate more fully on the remarkable use of a form of self-reference that goes beyond the attachment of labels. The 'this', being the dusty lorry, encompasses the dust itself, the very material of which the token is made. It is possible to go even further along this track.

Once I was driving behind another dusty vehicle and was so taken with another item finger-written in the dust that I nearly ran into the back of it when the driver braked sharply. It read simply, 'How's my spelin'?' The joke is of course that the answer to the question is embedded in the written words of which it is composed, giving a lovely image of innocent lack of insight. But the self-reference of the last word is deeper than a joke. There is the fact that the word 'spelling' itself has a spelling, so that it is possible to spell it incorrectly. There is the more interesting fact that the token is used not merely to signify itself but to signify a deviant form of itself, a deviation between the correct and incorrect forms of itself, and hence the innocent misspeller.

I tease apart these jokes, like an idiot pulling the wings off flies to unpack the secret of their flight, for a purpose: as a reminder of how human signs are riddled from top to toe with meta-language; how the many-layered consciousness that is captured in language is replicated in the many layers of linguistic self-consciousness or consciousness of language; and how the advent of writing has multiplied those layers.

Thi is relevant because I want to emphasize how self-referentiality is not a late entry into human signifying consciousness. There is an idea, very popular among literary critics and literary theorists in the final third of the last century, that there was something rather special about the self-referentiality of the postmodern novel; that it respected, as no other art form before it, and certainly no popular prose, the essential self-referentiality of language. This is not only wrong about the history of fiction but also wrong about *Homo significans*.[11] Readers will be familiar with the frequent threats issued by one cartoon character to another in the *Beano* to punch each other into next week's comic, a threat that illustrates the complex self-referentiality that is routine in popular prose, showing that it is not something that has been highlighted for the first time by postmodern fiction. More importantly, anyone who has thought about pointing, and, indeed, reflected upon the indexical awareness, in which it is rooted, will know that self-reference haunts human consciousness at pretty well every level above mere sentience. Except that we should be careful not to call it 'reference'. At any rate, indexicality reaches way down into human consciousness, as pointing *to* points *out*.

So, while pointing may seem like an innocent wildflower of primitive communication in a slick, tumultuous city of knowing discourse, it is nothing of the sort. The self-consciousnessness that haunts indexical awareness becomes more pronounced when we share our awareness through pointing. This self-consciousness permeates the act itself: pointing is riddled from the outset with meta-pointing; in short, implicitly at least, it points to itself. The pointing finger points to the pointing, if only because it is an explicit sign that cannot be anything else. It has to be, and look, like such a sign if it is to make its nature (and intention) clear and deliver its packet of sense. This self-pointing of pointing – and indeed of any deliberately engendered sign – opens on to the wonderful, dizzying world of token-reflexivity.

chapter seven
Assisted Pointing and Pointing by Proxy

Extending the Arm

The word 'pointer' has several meanings. It points, that is to say, in several directions at once. Another lexicographical pit stop is therefore indicated. The core meaning is, according to the *Oxford English Dictionary*, 'One who or that which points out'. A pointer may be: *a.* 'A person who points or indicates with his finger or otherwise' (this, the dictionary tells us is *rare* – though not in the present work);[1] or *b.* 'A rod used by a teacher or lecturer to point to what is delineated or written on a map, diagram, blackboard, or the like.' It is the second meaning that provides our cue in this chapter.

The topic of pointers is encyclopaedic. For pointers have moved on and travelled far beyond the classroom since the great lexicographer James Murray and his friends fought their way out of the endless 'web of words'. The reader need not fear a descent into inventory, for my aim is to unearth only some principles and to alight on certain examples in order to triangulate the territory. A further glance at unassisted bodily pointing is indicated before we examine the prostheses and proxies that *Homo indicans* requisitions to enhance his ability to point to, at or out.

We have already noted that the index finger (or the index finger-plus-arm) is not the only bodily structure humans use to point. We may tilt our head, explicitly direct our gaze or even point our foot in order to draw someone's attention to something in the shared vicinity. This is a striking illustration of the way in which the instrumental character of the hand, as the 'proto-tool', can infect the entire body, which can become a whole kit of tools. Off-hand, parenthetical, pointing – either metaphorically as when the thumb is used to indicate something, or literally, as when the head or lips or eyes or some non-canonical part of the body is employed – can be a means of expressive contempt for the individual seeking information. The use of the eyes as pointers is especially thought-provoking. You ask me where something is. My hands are fully occupied and so I glance in its direction. You make a critical remark about a mutual acquaintance, and I look with deliberate furtiveness to indicate to you that she is within earshot. Most of the work in these situations is done extra-ocularly. In the first case, my glance is the answer to a particular question that has defined the pointee: everything has been set up for the slightest indication to convey a key piece of information. In the second case, my glance simply indicates that there is something amiss and it is up to you to look around until the penny drops. Even so, ocular pointing is rather remarkable. It requires you to detect what are, at least from the usual speaking distance, minute movements, and to infer the location of the pointee from the direction of your gaze.

As with all pointing, successful communication depends upon the consumer placing himself at a notional 0,0,0 position from which the producer's glance originates. He occupies, as it were, the centre of her egocentric space. Artificial pointers deal with this challenge and aim, largely, to make it easier for the consumer. Without going overboard on pedantry, we may identify three sorts of pointers: those which augment the pointing part of the producer's body (the extended arm and index finger); those which go beyond augmentation to act prosthetically and cross the gap with minimal involvement of the arm; and proxy pointers that act in the absence of any particular producer. The augmenters and prostheses are 'assistive', the proxies are 'stand-alone'.

Before we deal with them separately, it is important to reflect on the common soil out of which they have grown: the explicit sense, discussed in the last chapter, that we human pointers have of what we are doing when we are pointing. The tool-like status of bits of the human body derives from the 'toolness' of the human hand that, as it were, infects the rest of the body with an explicit sense of agency. When we act, we know for the most part what, and why, we are doing so and this is particularly evident in the case of communicative acts, among them pointing. 'Meta-awareness' about pointing is evident in all societies, even in those which may be designated as 'primitive', such as the Arrernte aboriginals in northern Australia. This meta-awareness of the pointing finger (or head or eye or elbow or whatever) as a 'tool' – so that pointing in some sense points to itself – is a precondition for the use of non-bodily tools for pointing. At any rate, the use of a stick to enhance pointing is another reminder of the self-consciousness that pervades pointing. That self-consciousness is heightened in different ways, in particular by the retroactive effect of tools on the human body that make the body more aware of its 'toolness'. When we talk of our finger 'sticking out', we have a lovely example of a 'forecarted horse': the sticks we use as tools are used to characterize parts of the body. (It is one of the great conceptual achievements of humanity to see the human body as a piece of engineering – to approach it as a tool-kit.)

Let us begin with the archetypal pointer the *OED* spoke of: 'A rod used by a teacher or lecturer to point to what is delineated or written on a map, diagram, blackboard, or the like.' Many of us talk in our sleep. The distinctive achievement of lecturers is to talk in other people's sleep. In order to avoid this, and to make it easier for their students to understand what they are saying, lecturers enhance their lectures with visual aids that may be low-tech (such as chalk on blackboards) or hi-tech (such as PowerPoint).[2] Low-tech visual aids themselves require assistance, so that they can be connected with, and illuminate, the presentation, and this connection may be made with the help of a pointer. The lecturer points to the relevant part of the slide he is using in order to emphasize the point he is making.

He uses augmentation for two reasons. The first is to make his pointing more visible, given that he is at a distance from the audience. The finger sticks out from the body; the pointer is a stick that extends the natural stick. The second is to define the pointee more precisely. This is necessary not only because the latter may be quite small from where the audience is sitting, but also because the lecturer himself may be at a distance from his slides and needs some assistance in order to pick out the target datum, curve, line or pedagogic point. In recent years, there has been a revolution in didactic pointers: wood, metal and other solid means of enhancing the line connecting the producer with the pointee have been replaced first by light beams, and then laser light beams. The line connecting the producer with the pointee does not have to be underlined if, as happens with light pointers, the pointee can be fingered directly by light beams and the latter mark the pointee for our attention. The role of pointing as virtual touch or a tactile glance is taken literally: the pointee is touched by a proxy glance.

It is only a few years since those of us on the circuit gave up using lantern slides and converted to PowerPoint, and yet already my un-used slides seem like relics of a remote past. The pointer, too, seems likely to go the way of those little squares of plastic, for PowerPoint allows one to point electronically. The mouse drives an arrow across the screen and points unequivocally at the object of interest. Even the arrow is digested and the target may pick itself out, by lighting up (and so putting itself in italics), popping up as the newest comer on the screen, or flashing on and off (and thereby putting itself in double italics). This is the final stage in pointing, whereby the object says 'Look at me', rather as the bottle in *Alice in Wonderland* says 'Drink me'. The symbol on the slide is pointing to itself. Whether it thereby more effectively breaks into the audience's slumber than the age-old droning of the dominie unassisted by visual aids is un-certain. For many, the darkening of the auditorium at the onset of PowerPoint is the signal for the long-awaited PowerNap.

Towards Abstract Pointing

Pointing, of course, goes way beyond the lecture theatre, and travels in many other directions than into slides that point to themselves. Any examples will be arbitrarily chosen, but there is one that links with Michelangelo's Finger by a somewhat unusual route, and so this is the example I will choose. When cars were first introduced, there were so few vehicles on the road that it was unusual for one car to pass, or crash into, another. Soon the roads became sufficiently cluttered, and cars sufficiently fast moving, for it to be necessary for drivers to be able to predict what other drivers were about to do. Signalling of intentions became essential to road safety and, indeed, to survival. One of the most important signals related to changes of direction. In the first instance this was indicated manually: drivers and passengers pointed their arms to indicate the future direction of travel. This was in itself an interesting development since the pointee is not an object but a *direction*; more specifically, an *intended direction*, given that pointing has to take place sufficiently in advance of the change of direction to give other drivers time to adjust their own trajectories or planned trajectories.

There were two problems with manual signals. First, they required taking a hand off the wheel, which was a bad thing when cars often had to be actively steered even to maintain a straight line. Secondly, while it was comparatively straightforward for a driver on the right-hand side of the car to signal an intention to turn right, it was less easy to signal a left turn. There was a third problem: winding up and down the windows was a major distraction from driving. The case for a prosthetic pointer signalling changes of direction was manifest, and the indicator was born. These were little orange slats made of plastic, which lit up at night, and were fixed at about ear-level between the front-windows and the windshield. All that was necessary for the driver to do was to press a lever and the indicator on the appropriate side assumed the erect position, pointing in the intended direction of travel.

Every solution to a problem creates another problem situation, as the great philosopher Karl Popper pointed out. Apart from the pleasure that indicators gave to vandals – they were things to rip off, giving a satisfying snap as they parted from the car – they also had the habit of getting stuck in the erect position. In 1957, when I was nine, I was taken by my friend Roger and his parents for a week's caravanning in the Lake District. The journey from Liverpool took four hours, and it gave Roger and myself much subversive pleasure each time he announced the *double entendre*, 'Your indicator's up, Dad,' when the device in question jammed in the erect position. The link between the erect penis, indication and the index finger is of much etymological and psychological interest, but not sufficient to warrant more attention here.

The problems just described prompted the rise of the winker. The mental leap from an indicator modelled on the index finger to the winker is rather extraordinary.[3] The resolutely unimpressed might argue that the winker was still rooted in bodily pointing: it was a glance. But of course it was yet more abstract than the glance. The winker does not swivel. It draws attention to itself by alternating between its presence and its absence. (This fact is exploited in the joke about the – here readers supply their own group of individuals whom they regard as cognitively challenged – about the person stopped by the police and asked to see whether his/her rear winker is working. S/he inspects the light in question and, seeing it going on and off, reports: 'Yes it is. No it isn't. Yes it is. No it isn't...') At any rate, the design of the winker spares itself the necessity of spatial movement, replacing it with temporal movement. The translation of this occultation into a mode of pointing depends upon a code whereby the side of the car that is preferentially lit up becomes itself a pointer of the direction which the car is going to take: pointing is achieved through preferential siding to denote not a direction but an intended direction. Abstract or what?[4]

Stand-Alone Pointers

Which brings us to stand-alone pointers, pointers that point without producers, or, at any rate, continue to point long after they have been deserted by anyone's specific communicative intention. I am thinking about proxy pointers such as signposts. These are a manifestation of the extraordinary fact that humans can store information or, more precisely, the possibility of information outside of their bodies – in what Karl Popper called World 3, a realm of potentially informative artefacts additional to the world of material objects (World 1) and that of mental phenomena (World 2).[5] But there is something additionally strange about signposts and, indeed, Public Notices, of which they are a rather special sub-group.

They are curious, even eerie. As a child, I used to be haunted by the little pictures at the end of the chapters in adventure stories, which often took the form of a solitary signpost. Even now, when I pause by a signpost in a lonely place, I can sometimes feel that they have the ghost of intention hovering round them, which can translate, when one is in a hyper-suggestible state, into the intention of a ghost. The continuous possibility of informing that takes place in the absence of an informer or an informed is rather reminiscent of those wind-driven prayer-mills set up by lazy monks to do their work for them. Signposts are haunted by the absence of those two presences that are normally required to uphold what semiologists would call the 'deictic space of pointing': the body of the person or persons who erected it (and of those whose decisions and decrees authorized the erection); and the body of those who receive the information it makes available.

Signposts are haunted not by transcendent spirits, but by the collective spirit of the culture or society to which it belongs. We are addressed by the general other and are thus ourselves reinforced as generalized selves, as individuals who to some extent see themselves as (to use the term introduced by the founding father of social psychology, G. H. Mead) internalizations of a 'Generalized Other'. In being thus addressed, we are in some measure ourselves generalized, nudged in the direction of the condition Heidegger called *das Man*, or '*the they*', substitutable 'everyones' and 'anyones', atoms of the public to whom all signposts speak without fear or favour. As consumers of the information afforded by such signs, we are part of a bodiless body of folk who are spoken to by part of a bodiless body of folk. What Auden said so beautifully of public timepieces – 'And strangers were as brothers to his clocks'[6] – is just as true of signposts. They are a general assistance, anticipating your general need, the questions you might have in common with your fellow men. We read the sign, we wake it up, we make it speak for us, and the eerie feeling comes from the fact that our needs have been anticipated, and hence generalized, and so too are we. The promiscuous sign reveals its secrets to all, like Shakespeare's whore who doth 'unpack' her 'heart with words' to every customer who hires her.[7] There is a bleakness and an emptiness, as well as the comfort of familiarity and even community in the notices that address us in public spaces; that invite us to notice what may be to our benefit. Or command us to take notice – for notices are both nouns and verbs; objects and commands.

The numinous atmosphere is reinforced when the post takes the form of a *fingerpost*. The *OED* defines this as 'A post set up at the parting of roads, with one or more arms, often terminating in the shape of a finger, to indicate the direction of several roads; a guidepost,' while *Webster's* gives both 'a guidepost bearing one or more index fingers' and 'something that serves as a clue, indication, or aid to understanding or knowledge'.[8] Such posts point out directions by *materializing* them in a way analogous to the original carnal pointers which they imitate to a lesser or greater degree. The direction pointed out is a 'direction to X' where the pointee (X) often remains invisible: there is no purpose in looking along the arm of the fingerpost in the hope of seeing it. One has to take its point on trust: both the existence of the pointee and its location. Because they can indicate only a direction, and the pointee is invisible, fingerposts cheat: they have the name of the pointee and the distance from the pointer written on them.

This is a curious deviation from the primary grammar of pointing, not the least because it restricts the pointer itself to a single pointee. The name – 'London' – and the mileage instruct the consumer to overlook any other object, person or place that may happen to lie along the axis of indication. The pointee is thus a kind of internal accusative of the pointer. Securing this incestuous relationship requires an interesting variation on the activity already discussed, namely linking the space of indexical awareness (which is occupied by the visible fingerpost) with the space of deindexicalized awareness (which is occupied by its pointee as a possibility). The link is secured by the rather transgressive activity of placing the name of the pointee at a particular point in space: on the lath of the pointer.

When words are used to refer to objects, it doesn't usually matter where the actual tokens – the sounds or the letters – are located. 'Fido' means the dog in question whether I utter it upstairs, downstairs, in Manchester or in Paris. Indeed, this is linked with a fundamental feature of signs such as words that reach into the deindexicalized realm: their referents are not located with respect to the body of the person using them. There are, of course, very important exceptions, as we have already seen, when we rely upon the spatio-temporal context of an utterance to specify the particular example of some general type to which we are referring. It may be implicit (as when I refer to a dog and there is only one in the vicinity) or explicit (as when I use words such as 'that', 'this', 'here', 'now' and so on). This phenomenon of deixis is, as we noted in the chapter on ostension, a link between the locationless, general system of language and the world of located, particular things; it is how we can pick up particulars by means of generalities. This is not how it is with signposts. For, by definition, the singularity of the pointee is guaranteed. The 'London' indicated by the signpost is 'the' London. This is true even when there is more than one entity bearing that name. For example, a sign to 'Wellington' can be used to indicate a least a dozen places. Any given sign, however, will indicate the Wellington in the vicinity. The usual rule of naming towns is not to give two adjacent towns the same name. The development of more extended literal and mental maps has caused some trouble, but this is a late development. It is only recently that it has become necessary to specify that one is talking about, say, Wellington, Shropshire rather than Wellington, Somerset or Wellington, New Zealand. A more difficult case might seem to be a 'Way Out' sign, but here deixis comes to the rescue: the sign implies 'Way Out of the Place You are In'. And the place you are in must be next to this sign if you are looking at it. The word on the post is a token of an *essentially* proper name. The location of the word is crucial because what is being established is a spatial relationship between the token of the name and its object. The printed word is a thing among things. So there are two kinds of relations: the referential relationship between 'London' the word and London the place; and the spatial relationship between 'London' the material instance, the token printed on the signpost, as indicated by the direction of the arm of the signpost, and the town in question.

The thinginess of the word, or its realization as a written token at a particular point in space, is emphasized by another feature. The word is not infrequently printed along the lathe of the signpost. It thereby instantiates the direction of travel necessary to get from the pointer to the pointee. Even more engagingly, the gaze of the traveller passing from the beginning to the end of the token reproduces the trajectory linking his present position with his destination. The written word has a long axis and this becomes, not so much a finger, as a secondary pointer.[9] This is an interesting variation on the naive notion of a word as a label tied to an object. Here it is the sign that is labelled in order that its significate shall be specified.

Things are not always so precise and there are many interesting variations. Motorway travellers will be familiar with signs such as 'To the North' or (since the 'To' is implicit in the sign) 'The North'. The North seems to occupy quite a specific direction when you are far from it: a finger in London pointing to Newcastle would have to move perhaps only 20° of arc in order to point to Carlisle. As you get nearer to the North, it modulates to 'the northerly direction'; once you are in it, it is all around you and, while it is possible to point in a northerly direction, it is no longer possible to point to 'The North' without redefining it upwards, taking advantage of the fact that the direction is a relative concept, though it has positive realizations. Concerning the ubiquitous French road-signs, '*Toutes directions*', one can only marvel that, in two simple words, it is possible to point in all directions at once.

We perhaps take it too much for granted that we can point to locations, particularly huge and only vaguely bounded locations, such as 'the North' or 'London'. That we can indicate the relationship of a particular place – where the signpost is located – to a massive sprawl, many square miles in area, encompassing tens of thousands of streets, hundreds of thousands of buildings, and other civic amenities and spaces, and millions of people, is a remarkable example of the manner in which pointing, when working in conjunction with language, grants us a reach greater than ourselves and certainly greater than our grasp, and permits a virtual grasp a million-fold greater than any actual one.

We have noted that there is something abstract, extending beyond possible experience, about the entity the signpost points to. Indeed they partake of concepts, for in the end 'London' and 'The North' are bounded by stipulation rather than being naturally bounded objects. That which is gathered up in the word could not be traced by a fingertip or even by the imagination. And when we look at, think about, or point to any of these entities, we relate only to part of them, to a portion that counts as our 'London', or to a 'North' of the mind. The pointer that points to 'The North' points to something that lies between a place and a concept, to something that is between a part of semantic space and something that is spatially related to our current location. The pointer by its very nature points to a collective reality to which we belong and from which we have grown. The 'London' of the signpost is the London of *das Man*.

Pointing is rooted in the first instance in joint visual attention, which depends on our sharing a visual–spatial frame of reference with others. It brings together a sense of the independent existence of objects, an awareness of other viewpoints than our own, and the notion of a shared public world. It is under such circumstances that we individually become 'One'. This is a mixed blessing. Coming to terms with being a substitutable atom in an infinite collective, an instance of *das Man* as well an irreplaceable self, is part of the process of growing up and into the world. In the nursery and early childhood everything that is addressed to you is addressed to you alone. When Father Christmas sits you on his knee, the queue of which you have been a part vanishes. This charming egocentricity is captured, and mocked, in the lovely mixture of deixis and deindexicalized signification seen in the final frame of a *Beano* cartoon: the unfortunate character who has missed his bus hobbles along a winding, moonlit road past a sign that says 'Home 10 Miles'. To be located in a world that is anyone's world is the unique condition, curse and joy of humanity. 'If we could communicate with the mosquito, then we would learn that it floats through the air with the same self-importance [as humans], feeling within itself the flying centre of the world,' Nietzsche wrote.[10] In practice, that is precisely what we do *not* feel.

The little digits that are sometimes placed at the end of fingerposts seem to underline the presences that haunt them. They also highlight how signposts point from indexical into deindexicalized space, indicating the named invisible that is, however, thought of as having a definite location in relation to the present spot. This concrete relationship between a concrete location and an abstract idea is a kind of halfway house. No wonder fingerposts encourage such lovely metaphors as those cited in the *OED*:

So many finger-posts, pointing your thoughts, along various roads, to times and countries far away.

It had pleased him to christen the pronouns the finger-posts of language.[11]

chapter eight

The Transcendent Animal:
Pointing and the Beyond

If we are to think of man not as an organism but as a *human* being, we must give attention to the fact that man...has his being by pointing to what is.[1]

The World of the Pointing Animal

At first sight, pointing seems both obvious and trivial, worthy of the attention only of the kind of anorak who delights in the pedantic description of minutiae. By now I hope the reader recognizes that pointing occupies a central place in human communication: ubiquitous in our daily intercourse with others, it is the first, and by far the most common and most versatile sign the infant acquires, present several weeks before the first recognizable spoken word. It has extraordinary complexity, drawing on, and linking, a variety of spaces – the physical space anchored on the gesture itself, an interactional space established through joint (usually visual) attention and, in the case of pointing integrated with speech, a *narrative* space.

This complexity itself points back at the complexity of our human being. We have noted that, notwithstanding certain questionable exceptions, pointing is well-nigh universal in humans and, what is more, it is unique to humans. Human groups who do not utilize index-finger pointing are in a tiny minority, and individuals within them still make indicative gestures – deictic gestural references – with other parts of the body. Those animals who have been credited with pointing, namely chimpanzees, do not do so in the wild, and their apparent comprehension of pointing in captivity turns out to be due to a rather over-eager application of the argument by analogy. Chimpanzees, what is more, do not point for each other's benefit nor do they encourage their young to point. The notion of pointing as referring to an object, and a mode of meant meaning, lies beyond the horizon of chimpanzee consciousness.

The fact that pointing is both universal in, and unique to, humans should alone make it worthy of study. It immediately suggests that it is rooted in something deep and distinctive in human consciousness. It was this intuition that attracted my interest to the gesture thirty-five years ago. The penny dropped – I saw the point of pointing – only after I had written extensively about human consciousness, which enabled me to see how much, drawn from a distinctly human awareness, genuine referential pointing required, most notably the sense of one's self as an embodied subject, and hence as an agent, and of others as selves with viewpoints, different from one's own, trained on a shared public space. In order to want to point, we need to have a sense of other people being endowed with inner mental states and of objects having an existence that is independent of one's own experiences of them. Pointing implicitly points up a common human space. Linked with this is the notion of a continuing hidden reality that underpins and surrounds the bubble of sensory experience in which one is situated, a reality that is hugely expanded by manufactured signs. To put this another way, an individual who points dwells in a *world* rather than being located in a mere environment. A world is a boundless sphere of explicitly shared possibility, not merely an array of material objects encircling the material object that is an organism.

Pointing both grows out of such a world and helps to build it. This dialectic or iterative cycle is universal in the process of world-creation and spans a boundless multitude of people living together, or only remotely connected or even separated in time. The general principle that tools make us human just as much as humans make tools is particularly true of the primordial, most immediate, tool, the hand, and its digital servant, the index finger. We see co-evolution of shared, or common, signs and objects and shared and individual consciousness. The mode of consciousness which is a prerequisite for us to become pointing animals is itself enhanced by our indicative behaviour. Pointing seems such a minute gesture and the primordial pointer such an unimpressive structure, but it cements and develops the joint attention from which it grows; and the cumulative effect of shared indications helps to stitch the great space that is the human world.

In this final chapter I want to investigate the link between pointing and transcendence, from its most ordinary manifestations through the sense of the transcendent realm which is discussed in religious discourse. The transcendent, which is the most salient condition of our being able to point, and which is then enhanced by pointing, is rooted in the intuition of the hidden, in the presence or reality of that which is the unobserved, absent, beyond. The intuition of the hidden lies at the heart of man, the Pointing Animal – and indeed, man, the Knowing or Explicit Animal. It is this – transcendence, the present of the absent that pointing, ultimately, points to.

Touching and Seeing

Let me step back a little and begin with touching. Close your eyes and reach out for an object, let us say a cup. You feel its surface, its resistance, its weight, its shape, its size. You have, that is to say, a multitude of sensations located on the palm of your hand and your fingers. These bodily sensations do not simply report the state of your hand: *they reach out beyond themselves*. They disclose, or seem to disclose (for illusions, as we know only too well, are possible), an object, an object other than themselves: the cup in question. This fundamental phenomenon, whereby one material object (the human body) discloses another, is deeply mysterious. We have no way of explaining it, though many philosophers have persuaded themselves by means of linguistic prestidigitation that they know how it happens: that it is down to the activity of the brain. I beg to differ and have done so at great length. I won't repeat my arguments here, but just focus on this amazing phenomenon, whereby *one object* (in this case my hand) *discloses another* (in this case a cup).

The sensations you feel in your hand are *about* something other than themselves. Philosophers call this property of 'aboutness' *intentionality*. You are led to believe in the existence of something that is other than the sensations that you are feeling. The sensations are in one sense about the hand. Yes, they are experienced as sensations in and of the hand; but they are also about the cup. They are the means by which the tactile surface of the hand reports that the hand is holding something. Because they are *about* something other than themselves, the sensations form the basis of a perception. Perceptions, unlike sensations, can be wrong. I cannot think incorrectly that I am having a sensation, but I can think incorrectly that I am having the sensation *of* something, particularly something of such-

and-such a nature. What I think is a cup may not turn out to be a cup at all. Indeed, I may even be deceived that the sensations that I have are of any object or even of my hand. Patients who have had amputations may experience sensations in a hand they no longer have. In virtue of their intentionality, sensations lead us to propose the existence of objects that not only transcend the sensations but also the hand and ultimately the person whose hand it is.

The word 'transcendence' is typically associated with something rather more up market, the condition, for example, of an elevated being, such as the God who made Adam, who inhabits a hidden world, who lives above and beyond his creation, answerable to no laws, natural or any other kind. For an atheist like me, however, there is another meaning of transcendence, which is remote from the divine, though, as I shall argue presently, it lies at the origin of the notion of the divine. It is the homely, everyday – indeed every moment – transcendence of ordinary perception. Let us look at the transcendence of the perceived object a little more closely, through the eyes of Phenomenology, that great school of twentieth-century philosophy headed up by Edmund Husserl, Martin Heidegger's teacher.

Pick up that object again and think about it. It is obvious that you grant it all sorts of properties that go beyond the ones you can presently feel. For example, it has layers beneath the glazing, lying beyond your touch. What is more, you are confident that, when you are no longer holding it, or perceiving it at all, it is still there, existing in its own right, independent of you. This is the sense in which the object is transcendent: it has hidden properties and, what is more, it can never be completely disclosed, never mind exhausted, by your experiences of it. You might imagine these unsensed properties –

indeed you are doing it all the time. Perception always goes beyond what is sensed to what is unsensed: the visible object to the parts of it that are not visible: hidden surfaces, tangible properties, interiors and so on. When we perceive an object, or infer it from a sensation, we experience something that goes beyond what is currently appearing to us. This is transcendence in the sense that I want to use it just now.

Our previous experience deepens the hinterland of the unsensed beyond what is appearing to us. When we perceive an object, we perceive it as an example of some general class or type of objects, or we perceive the material of which it is made as being of a certain kind. Expectations are awoken – *unconscious inferences*, the nineteenth-century physicist and physiologist Hermann von Helmholtz called them. They are unconscious in the sense that we may become aware of them only when they are overturned. You open your eyes and, lo and behold, what you see is not something with a hollow inside but a solid cup-shaped object. The most important thing about the expectations is that they are of a *general* nature. We have a general idea of what we might find if we explored the object further, a general idea of what the object would do under different circumstances and, more narrowly and practically, a general idea of what uses to which it might be put.

All of this – intentionality, transcendence, and general classification with general expectations, *a general sense of possibility* – is present in touch. What is given to us in tactile experience goes beyond what we are currently experiencing: touch puts us in touch with more than we are currently touching. This is true to a much greater degree in vision. For many reasons – the upright position, the special anatomy of the eye, the relationship between the eye and the brain – vision is uniquely well developed in humans. We are not talking about straightforward things like visual acuity – we could not compete in this respect with birds of prey – or even the width of the visual field. We are talking about vision as a means of making sense of the world. The most important feature of vision, from the viewpoint of our present concerns, is that it reveals the unrevealed. Of all the senses, it is the one that most explicitly has *a field*. The field is continuous: we not only see objects, we also see the spaces between them; we not only see things, we also see the distance at which see them. Most importantly, we literally see that there is more to be seen of objects than we are currently seeing: we see that the visible surfaces conceal invisible depths or conceal surfaces that are not revealed to us. The invisible is not merely implicit as 'the more to be seen' (compare this to 'the more to be touched' that we intuit when we touch an object) but explicit as that which is *visibly concealed*: we *see* that the object behind another object is partly hidden by it; we see that an opaque object has an invisible (but potentially visible) interior. The invisible interior is visibly invisible, while the untouched surfaces of the object are not tangibly untouched. Indeed, a degree of opacity – visible invisibility – is a necessary condition of visibility. An object that had no invisible interior and did not block our view of other objects because it was entirely transparent would be invisible. There is no equivalent in the case of touch: we do not have a tangible tactile field between the things we are touching or have a touch of the intangible. Nor is there in the case of hearing: sounds are not linked by a continuous field of sound, and we do not hear the inaudible. How different this is from vision, where all the visibilia in a visual field exist in an instantaneous relationship with one another and form a continuum of the visible and the visibly invisible, of what we can see and what we can see that we cannot see. The luminous is truly numinous.

The seeing eye sees not only objects but also the distances at which they are seen, or, more generally, the condition (background illumination, clarity and so on) in which they are seen. The gaze looks out beyond the body. Hearing is also a distance receptor but it is a receiver and does not reach out in quite the way that vision does. This is a profound dissimilarity between the two senses. The eye makes unmediated contact with the object: seeing is a kind of virtual touching, which explains why it is possible to feel groped by another's gaze. The transcendence of the object as something other than one's experience of it is made explicit as one crosses the gap between one's self and it. The virtual touching that comes with visual attention is the precursor to the virtual touching of the act of pointing. There is another bridge to be crossed before we reach true reference – the linguistic touching that 'brings back' the object. The importance of vision is that it takes us to the beginning of that bridge, and pointing moves us further along the bridge.

Let us look further at this distinctive aspect of vision. In *De Anima*, Aristotle said of perception that it 'gives us the form of an object without its matter'.[2] This is a deeply thought-provoking claim and there are different ways of interpreting it. But there is one interpretation that seems to apply particularly to visual, as opposed to other modes of, perception. When I see that object over there, I see its outline, or, rather, one of its outlines. And I see its surfaces. The material content – with its weight and resistance to penetration – is inferred, rather than seen. What is more, in vision we see the object and the field of objects surrounding it as a whole, all at once – that's why we see it as a form. Touch, by contrast, gives us the object and its neighbouring objects bit by bit, sequentially, and by direct encounter with its material. And the other senses give us the object only indirectly, through inferences, trained in part by sight and touch: we learn to 'hear', 'smell' and 'taste' objects. This abstraction of form from content, most clearly and fully and explicitly developed in vision, lies at the heart of classification: we see the object *as* a cup, or whatever. Sight is the key to classification and to generalization.

This, as we shall see in a moment, is intimately connected with the crucial role pointing has played in the development of human consciousness; but let us pause for a moment to consider classification and, in particular, to rescue it from the jaws of behaviourism. For a long time, there has been a fashion to reduce, or assimilate, classification to discriminative behaviour. If I behave in a certain way towards a particular type of object and another way to another type of object, it is argued, I have effectively established, through my behaviour, two classes of objects. To classify something is simply to behave towards it in a certain way. This account of classification allows us to ascribe classifying behaviour to all sorts of beasts, even quite lowly ones, such as squid. Behaviourist reduction of classification, however, overlooks many aspects of classification, such as, for example, *explicitly* assigning an object to a class, often applying a label to it, stating what its class is. The most obvious flaw in the identification of classification with discriminative behaviour is obvious to anyone who is not ideologically committed to behaviourism: namely, that we often classify objects without reacting to them in a particular way. Indeed, of the rich clutter of objects around me that I assign to different classes, very few elicit from me a particular pattern of behaviour. The second flaw, and almost as obvious, is the fact that we not only classify objects but also re-classify them. Our classifications are explicit and up for discussion.

There is, in other words, a distance between the classes to which we assign objects and the behaviour that follows from this. Very little necessarily follows behaviourally from my classifying an object as a 'cat'. I may or may not stroke it, feed it, avoid it, tease it, call out its name and so on. The argument that *circumstances* dictate my behaviour – that classification plus circumstances will together deliver a particular mode of behaviour – will run into the difficulty that most circumstances are unique and difficult to characterize in such a way as one can derive a one-to-one map of behaviour. More importantly, the general category 'cat' rises above individual circumstances or types of circumstances, so the class 'cat' cannot be assimilated into a range of possible behaviours. More specifically, a class isn't a cause, or a stimulus; or a type of cause or a type of stimulus.

The classes to which we assign things, then, are not wired into types of behaviour. We have distanced, explicit generality, and this is what it is to have perceptions that give us the form of objects liberated from their matter. Vision makes objects explicit, as existing in themselves out there, with potential general properties and hence amenable to classification. In the case of a seen object, however, the generality is still rooted in a particular. The cup that I can see is an instance of a general class but the general class is instantiated in a particular object. The cup is both general and particular. This hybrid status was captured by Aristotle when he talked about certain objects being 'sensible' (that is to say, accessible to the senses) and 'intelligible' (accessible to thought). What will take us decisively from the realm of the sensible (particular) to the intelligible (general)? The reader will not by now be surprised to discover what I am leading up to: that pointing has a key role in opening up a realm of pure possibility, of generality, beyond that of sense experience.

Pointing and the Human World

Let us return to the basic situation we discussed in Chapter 2. 'A' is on the ground while 'B' is at a vantage point, up a tree. B sees something of importance to both A and B. Imagine we are somewhere in that 200,000-year period between the emergence of *H. Sapiens sapiens* and the origin of language. B points at the object and makes warning grunts. The object of B's concern is imported into A's consciousness. A is aware of the object, but its presence in his consciousness is entirely general. It does not, as far as A is concerned, have a definite location – at best, it is somewhere 'over there' – and it does not have definite features. In Aristotelian terms, the pointee is an intelligible item that is not (yet) sensible, occupying a place in a world of meaning, uprooted from that of sensation. For the object I am pointing out to you, or pointing towards for your sake, and which you cannot see, is not precisely located so far as you are concerned – it is somewhere in my pointing field and may be nowhere at all if I am deceiving you – and has only the most general characteristics. The as yet unseen pointee, which is known only as the object of someone else's sight, is an intelligible liberated from being a 'sensible'.

The 'that' in the '[Look at] "that!"' of pointing is on the threshold of the weightless 'thattering' of discourse – asserting *that* such-and-such is the case – that fills our lives wall-to-wall. While to see stand-alone pointing as a telescoped assertion would, for reasons we have already discussed in Chapter 5, be to impute too much to it, it is certainly an 'Ur-that', which opens the way to all the different modes of 'thattering' that the Explicit Animal engages in. It brings us to the edge of the great world opened up by discourse. With pointing we have moved decisively from the solitary material world of non-human sentient creatures to the shared, public world of humans; from the organism that feels warm or cold to the person who feels, and subsequently asserts, *that* it is warm or *that* it is cold. Pointing is predicated on an uncoupling of the individual from his or her individual material environment, developing a gap which is then crossed by explicit signs, a gap which eventually will enable what is there to be referred to, to be asserted, a space in which truth and falsehood can arise. This world does not privilege me as its centre. More precisely, the world of pointers and pointees has more than one centre, from which it follows that it has no master-centre except inasmuch as my seeing you as someone to whom I wish to point something out locates you in my visual field and I in part recover the world as centred on my body. Henceforth, however, my sense of being at the centre of the world revealed to, and concealed from, me is haunted by the sense that this world has other centres. This is the first step towards understanding that the world is a public space that has no intrinsic centre, and that the privileged position I occupy in it is an illusion and the first inkling of the fact that there is an objective truth of the world which is not only what is revealed to 'a view from nowhere' (to use the contemporary American philosopher Thomas Nagel's famous phrase) but also 'a view from no one'. With pointing, 'That X is the case' begins to be liberated from *my* existence. Pointing stands on the borderline between enhanced indexical awareness and deindexicalized awareness, where that of which I am aware is no longer related to my body or, indeed, to my existence, that existence which the existential intuition has made explicit.

This step towards generality builds upon vision. There is the obvious sense that none of us could participate in the pointing game without vision. It requires the reciprocity of vision: I see you pointing and you see me as a suitable recipient of pointing. It might be argued that hearing might provide the same. Eventually it will, in the form of spoken language, but something has to be established first, and that depends on sight. For vision has provided us with the intuition of the hidden and the general, and also with a precursor to the virtual touching or virtual grasping of indication. Pointing builds on these, in particular on the visibly invisible or the invisibly visible. Most importantly, through pointing to an object that one person (A) can see and the other (B) cannot, we have an object that is presented to A as a particular and to B in a general form. The form of the object truly is presented to B as a form without matter. And it is in this that the importance of pointing lies.

Pointing has opened up a new kind of space: a space of possibility, beyond the sensory field, inhabited by objects that are presented in a general form. While B may anticipate what sort of object A is pointing to, his anticipation, even if 'good enough', leaves many features undecided. This interchange between A and B makes explicit the fact that the sensory field itself is located in a larger field of reality and possible awareness. Eventually this will grow into a field of collective awareness – which amounts to a shared world – that will subsequently expand as we communicate ever more sophisticatedly and pool our experiences. This will be a world of 'generalized "everyone's" experiences' that we may call knowledge.

Let us look again at the notion of the transcendent. It has several characteristics but two seem to me worth reiterating: it has the form of a *possibility*, which is of a general character and thus approximates what is ultimately captured in words (which assert, deny, etc. general possibilities); and it is *public*, that is to say it is equally and explicitly available to all. The notion of reality that is available to others as much as one's self (and may be available to others when it is not available to one's self and vice versa) is the lynchpin of the human condition of living in an acknowledged shared world. It is the foundation stone of a genuine social existence that is rooted in a sense of having a common arena for potentially shared experience. Other animals, such as so-called 'social' insects, may have dovetailing automaticities, but they are not truly social, for their interactions are largely prescribed and instinctive. They interact without this being deliberate, intended, willed or understood by the participants. Humans, by contrast, interact on an illuminated stage that is built up out of mutual acknowledgement. They have a genuine common space made of intersubjectivity. This public space is not merely the physical space in which the human organism is located, but a beyond, which is distanced from, and has characteristics different from, the space in which all non-human creatures interact with the physical world and, via interlocking mechanisms, with each other. There are many expressions of this: artefacts, artefactscapes such as cities, institutions, laws and principles, modes of discourse and so on. They constitute a theatre of deliberate behaviour, of actions that could not happen without the actors being aware of their purpose, of a world of possibilities which they intuit and try to shape.

The most elementary building-block of this shared space, requisitioned with each new entrant, each developing infant, is the interaction of the gaze, where each sees that s/he is seen by others and that the other sees. We are unique, as Daniel Povinelli has demonstrated, in having a sense of the other person as seeing, as having a viewpoint. When we see that another sees, this visible seeing reveals another kind of hiddenness: the psychological state of another person (so-called second-order intentionality) and the possibility of a different or parallel world as revealed to another's gaze. The psychological state of another person is a deeper interior, a more esoteric 'darkness visible', than the interior of an object or the objects round the corner. Pointing grows out of this joint attention into a shared visual attention to objects that build into a shared world, a public reality, an interpersonal arena for true society. By this means it inflates the bubble of available common reality in which humans operate. We are not (as are other animals) sentient monads interacting with, but inwardly sealed off from, each other, solipsists in practice without the notion. We are participants in a collective consciousness.

Of course, in the developing infant, pointing is very soon over-shadowed by language in the narrow sense of speech, and is caught up in the world that words articulate. For this reason its significance is overlooked. The present book is an act of affirmative action for a gesture of supreme importance, of itself and for what it reminds us of our own nature. If one subscribes to the idea that language origi-nated out of gesture, then pointing, as the most versatile of all gestures, and the one that seems closest to the primary, that is to say the referential, function of language, we may argue that it is crucial to the beginning of truly social being (of a kind unknown elsewhere in the animal kingdom), of a collectivization of consciousness upon which community, discourse, civilization and knowledge are based. The reinforcement of joint visual attention by pointing, and the invi-tation to another to attend to something of which she was unaware, and checking that she is aware of it (as infants do from very early on), or whose significance she had not appreciated, or which lies outside her current sensory field, is a huge step in affirming a shared world and a world which includes things of which we have no understanding, which provoke us to further our understanding. It is from this that arises the unique power that we humans have to dominate the natural world and, indeed, to create a parallel human world of artefacts and laws that are additional to, or displace, or are at odds with, the laws, processes and products of nature. We put our minds and bodies together to have effects that dwarf even earthquakes. Our collective experience and the power drawing on this is hugely magnified beyond what is made available to us by our bodies.

Pointing is a fundamental action of world-sharing, of making a world-in-common. It not only tacks that common world together; it also expands it, and the two processes are not separate. When you are pointing to an object that I cannot see, and which is actually beyond the horizon of my visual field, you are affirming the existence of a world lying outside of what I can sense. You are affirming possibility beyond actuality by pointing to a particular possibility that I cannot see. The boundless 'Beyond' is what lies through the doorway leading from indexical to deindexicalized awareness, glimpsed through the oriel opened up by your index finger. Of course, we are aware of things (events, states of affairs) that are not there before us – possibilities no longer realized, possibilities yet to be realized – without the assistance of pointing: we recall and we anticipate. What is special about pointing is that pointees are not only explicit but also *shared* possibilities, or, rather, what is actual for one person is proposed as a possibility for another. Pointing is a fundamental instrument in the socialization of possibility.

It makes the beyond explicit and makes it something shared.

The power of indicative gestures to awaken this sense of shared possibility is often compelling. A man standing in the road pointing at the sky will soon attract a crowd wanting to see what it is he is pointing at. This prank – and it usually is a prank – exploits the fact that pointing is, in the overwhelming majority of cases, to something that exists and which, moreover, the producer has reason to believe will be of interest, in particular to the consumer. He may believe this because it is of general interest to most people, or it may be of very particular interest to the consumer in question.

The possible, or the beyond, in which pointees lie has two dimensions, temporal and spatial: there are things that lie out of sensory range but are current; and things that are not-yet or no-longer. It seems that the out-of-sight (or out-of-sense) is the fundamental mode of the beyond: its 'not-yet-ness' or 'no-longer-ness' is set out in space. We have the sense of something coming (an emergent possibility) or of something going (a lost possibility) inscribed in the passage of things towards us or away from us. Pointing, which points to possibility, reaches into both dimensions at once. When you point to something you can currently see and I cannot, the pointee is located not only in a spatial beyond but, so far as I am concerned, in a temporal beyond. You are pointing to a future experience I might have if I follow your instructions to attend. With the assistance of language, of course, pointing may be more explicitly tensed, as when I point to an empty chair that was once occupied by someone who is no longer. The great emblem of Communist countries – that of the leader pointing towards the future – is extraordinary: the pointee is an entire (happier, better) state of a nation, to be arrived at, and located in, a non-specific future. Indeed, this is so general that the act almost turns in on itself and simply becomes a sign of itself – or of the kind of country in which charismatic, autocratic, brutal leaders remain in power by pointing to a better future and liquidating those who do not share their vision or buy their rhetoric.

The Boyhood of Raleigh

The promise of pointing – the future experience of realized possibilities – is often very complex and the consequences of their fulfilment tumultuous. One of the most striking examples is captured in a picture, Sir John Everett Millais' *The Boyhood of Raleigh*, well known to many children of my generation. It was intended to inspire us and to make us think about our past, our heritage, and the Empire. Indeed, the present meditation on pointing could as well have been titled 'Raleigh's Boyhood and Mine', but this might have conveyed an unfortunate impression. Given that I have written no histories of the world, led no military nor exploratory expeditions, been the favourite of no queen and the target of no king's lethal enmity, have not discovered the potato for the Old World, nor assisted the dissemination of tobacco, nor inflamed the imagination of others with the dream of discovering an undiscoverable El Dorado, nor yet had my head judicially separated from the rest of my body, there cannot be any question as to which of us is the greater figure. Besides, there is an inevitable asymmetry in our relationship: while he could be an influence on me, I could not be an influence upon him, except in the somewhat contrived sense of my having an infinitesimal effect upon his posthumous memory by invoking him as I do here.

There are, however, important connections between us. Raleigh's central role in the transfer of potato culture and his peripheral role in the transfer of smoking from the New World to the Old, has influenced the school dinners that I ate, the Woodbines that made me sick behind the outhouses, and the crisps and chips I sometimes consumed on the way home. But the connection I want to highlight is deeper and more important than this. It points to something rather fundamental about the *beyond* into which pointing points, the reach of its transcendence.

The Boyhood of Raleigh was hung at the back of the classroom that I occupied in 1955 as a first former just out of kindergarten. Young Walter is dressed in best Elizabethan garb: velvet doublet and ruff and expensive buckled shoes. Next to him is a similarly accoutred lad, perhaps a younger brother. Walter is sitting on the ground, near the end of a harbour wall, with his knees drawn up, his arms round his knees, and his legs crossed. He is leaning forward, wide-eyed, with a slightly anxious, but also fascinated, look on his face. His expensive hat, velvet and be-feathered, lies on the ground, quite forgotten. His brother, perhaps a shade more sceptical, is lying on his tummy, his chin supported on his hands. Notwithstanding their high-born status, they are listening with respect, indeed with manifestly rapt attention, to a floppy-hatted sailor in red pantaloons, a loose beige shirt with neck open to the ocean air and rolled-up sleeves exposing mighty forearms, and bare feet. He has a large, dark moustache, curly hair and a hoop ring through the ear that is visible to the onlooker. The legend under the picture informs us that the two boys are listening 'to tales of wonders on sea and land' told by 'a sunburnt stalwart Genoese sailor'. What Millais depicts is a defining episode in the history of one little boy. Since the boy was subsequently to become an icon of the adventurous spirit of England, buccaneering along the border between mercantilism and piracy, that moment has been chosen to stand for a defining moment in the history of England and of the Empire that was to emerge a few centuries later.

The sailor is pointing out to an open sea, a few shades darker than the eggshell-blue sky. What, I always wondered, would young Raleigh have seen if he had gazed down the long axis linking the sailor's index finger to an horizon many painted sea-miles away? What was the sailor – multiply framed by the pitted wood border around the picture, the cream-washed often sunlit (this was my childhood after all) wall, the borderland between my childhood and my boyhood, and the 1950s – what was he pointing to? The answer to the two questions may not be the same.

The legend would have us believe that Raleigh saw a land populated with exotica beyond the ocean's downward curve to invisibility. He saw wealth, fame and power. He saw a paradise of macaws, gold and glory. He did not see the malarial fevers that made El Dorado Helldorado, the plotting and jealousy that conspired with the flaws in his own character to ruin him; he did not envisage his own de-bodied head bleeding in a basket. He saw hope and never thought of danger. He was, in short, a true-born Englishman. He was, anyway, the very model of the merchant–adventurer, the paradigm of those early globalizers, those traders whose gains – well-gotten by the standards of the day and ill-gotten by standards we can now afford – enriched their country. These riches eventually (and how many hours are compressed in that 'eventually') fed into the disposable income of willing patrons and reluctant taxpayers that underpinned the places of learning and inquiry, of education and research, from which the technological revolution was finally born.

The transformative effect of the dreams of Raleigh and his like was noted by the great economist John Maynard Keynes:

The modern age [marked by unprecedented progress and invention] opened, I think, with the accumulation of capital which began in the sixteenth century. I believe…that this was initially due to the rise of prices, and the profits to which that led, which resulted from the treasure of gold and silver which Spain brought from the New World into the Old. From that time until today the power of accumulation by compound interest, which seems to have been sleeping for many generations, was reborn and renewed its strength. And the power of compound interest over two hundred years is such as to stagger the imagination…I trace the beginnings of British foreign investment to the treasure which Drake stole from Spain in 1580…Thus every £1 which Drake brought home in 1580 has now [1930] become £100,000.[3]

The impact went beyond Britain. As the philosopher Barry Allen has said, in the sixteenth century, 'things began to change. Treasuries were richer than ever. Its commerce and industry finally outpaced all the others and, with the plunder of the New World, Europe became the first center of the first truly global economy.'[4] The ripples spread beyond finance, transforming the nature of the relationship between monarch and subjects, most importantly widening the responsibility of the government to the governed. Hitherto, it had not been 'the king's business to take care of people' but simply to 'preserve dominion over territory'. The task of government was 'reconceived as the management of a complex resource: the population, its wealth, its health, economic and military potential...By the end of the twentieth century, the subjects of modern government are... accustomed to comprehensive care by professional agencies presumed to act with expert knowledge and benevolent intentions'.[5]

But these wider effects depended ultimately upon the free disposable resources appropriated by these merchant–pirate–adventurers. Money created leisure, leisure created science and science empowered technology, and technology lay at the root of the Industrial Revolution. Machines drew labour out of the cottage and the shop and amassed it in factories. Production aggregated to mass-production and mass-production, in addition to the suffering and exploitation and the immemorial characteristics of collective human enterprises that it brought, uniquely made possible philanthropy, benefaction, enlightened social policies such as universal education and, finally, the wider welfare-state. The fall in unit-cost made commitment to universal education non-utopian. Science, industry, a leisured class, entrepreneurs with sufficient income to endow this, that and the other, reformers, visionaries and progressive parliamentarians, all interacted in various ways over the nineteenth and twentieth centuries to create the world of my mid-twentieth-century childhood in which even the children of the working class – and only a generation separated me from the labouring classes – were educated in schools that were able to afford mass-reproduced pictures to hang on the wall. The exotic, unimaginable land beyond the curve of the sea, beyond the horizon, that the sailor was pointing to reached, across centuries of unimaginable change, to the world of my childhood. Unknown to himself, that stalwart Genoese sailor was pointing to my classroom, to the picture itself, to me, the astonished writer at the proximal end of the fingers typing these words. The exotic, wild land beyond the tip of his finger was the tame, domestic future that I inhabit.

The God Finger

It may seem absurd to suggest that globalization – the stitching together of the world – technological advance, the transformation of the narrowly prescribed role of the monarch into that of a government that permeates every aspect of our lives, grew out of the tip of an index finger. I have dwelt on this example, not only for the egocentric pleasure of pointing to myself, but because it illustrates the profound relationship between pointing and the sense of possibility that is a necessary condition of the transformation of the condition of hominid from that of a life merely lived to one that is actively led. And the horizon opened up by pointing takes us beyond this world, beyond even an ever-increasing distance from nature and material prosperity. The index finger points to another kind of beyond, the one that we began with, when we discussed Michelangelo's *Creation of Adam*. Let us take the final step in the journey from the everyday transcendence of perception to the transcendence of the divine, a journey that began with the invisible inside the visual field; continued with the hidden round the corner and over the hill; went beyond the far horizon; rose to the sky, that which was above the clouds and beyond the stars; and then passed away completely beyond sense experience – to the ultimate Hidden – the idea of God who sustains our days and underpins all that is, the Creator hidden behind his Creation.

It is scarcely surprising that man, the Pointing Animal, should form an idea of that which is ultimately hidden, ultimately general, the origin of all possibility, of that which lies on the far side of the space in which we have our shared lives, the human world. Given that we share attention, as well as resources and labour, it is equally unsurprising that this idea should be elaborated to an extraordinary degree; that we should have collectively developed a folded notion of the Hidden One, that from which the future comes, that to which the past goes, and that there should be a close relationship between this Hidden One and the ultimate end to the field of consciousness, the intractably hidden realm to which we depart when we go out of the sight of men, into the sunless zone beyond life. We point to the sky and see not emptiness but another life where meaning and being merge in significance as solid as a brick wall. In the light of this intuition, our life becomes 'This Life', hollowed by rumours of an afterlife in which we are complete, and where inexhaustible possibility is not at war with actuality.

The realm of the divine is defined almost entirely by hiddenness: it is this that classifies it as purely transcendent. This was made explicit by Plato, for whom the realm of ultimate reality and truth was inaccessible to the senses and available only to the intelligence. Aristotle, of course, opposed this and their disagreement is captured beautifully in Raphael's *School of Athens*. The opposing positions of the two philosophers are signified by two indicative gestures: Aristotle gesturing towards the earth and the world of sensible experience, and Plato pointing up to a heaven of pure intelligence. (In Jacques-Louis David's *The Death of Socrates*, Plato's teacher is pointing upwards as he is proffered the cup of hemlock.) The purer the vision of God, the more empty He is of characteristics. The Old Testament God, interfering in the affairs of men, laying down rules of conduct, behaving in such as way as to warrant an anger-management class, is an appropriation of the sense of the ultimate boundary of the hidden for political and other social purposes. This is in part because His transcendence is the collective awareness of the world that is society, that gathers up the hiddenness of so many viewpoints other than my own: it is inescapably contaminated. And yet, as that which pointing ultimately points to, He should have no qualities, being defined 'apophatically', or negatively. This has been brilliantly captured by Jean-Yves Lacoste, a theologian influenced by phenomenology:

> No 'God' is at our disposal, except the Idol. And the fact that the heart may feel both God's presence and God's absence, and always a greater absence, is the cornerstone of all hermeneutics of so-called 'religious experiences'…God's presence, in so far as 'presence' is understood as 'presence to the heart', is essentially frustrating…God's phenomenality cannot be understood if we do not understand that God transcends his phenomena.[6]

The longing for God has been described as 'the absolutization of desire', or of dissatisfaction. Such states are awoken with transcendence, which is the presence of the absent; ultimate transcendence is the presence of the ultimate absence.

This is unbearable, and so the desiring heart is always prone to false prophets who know the mind of God and can create verbal or non-verbal graven images. Such prophets tell us that God Himself has a finger and, through the idea of this forefinger, mankind's collective pointing turns back on itself, and through God's appointed agents, points at individual humans and human races, who are singled out for glory or execration. We are pointed out, beckoned and summoned to judgement when we die.

This is the final expression of the transcendence, and the sense of a transcendent realm, which is opened up by pointing. Pointing, which begins with the body, opens up spaces quite other than those by which the body, in common with every other object, is surrounded. But sooner or later it turns back on itself and via how-soever many long loops it reaches to the place where it began, though it is irreversibly changed by the journey it has had through the beyond.

Deindexicalized awareness has to be cashed in the here-and-now, in something that feels itself to be here-and-now rather than nowhere-in-particular at no-particular-time. At the end, it returns to the place from which pointing liberates us, where to Yeats 'all the ladders start'. Not his 'foul rag-and-bone shop of the heart'[7] – that seems too harsh; rather the warm ossuary of our living body. I recently sat by someone's bedside as she was dying. She was no longer able to speak: even the inarticulate muttering had come to an end. However, from time to time she seemed to point: to reach out from this body that was closing in on itself into a world of transcendence that was gradually shrinking. She was pointing out the end of pointing.

So long as we do not end, there is no limit to the beyond pointed out by pointing. The beyond that we move towards, that we conceive, is turned by our arrival to a hither haunted by another beyond. For animals, who do not get the point of pointing, the beyond is beyond them. We humans, by contrast, are never free of its haunting, its present–absence: the invisible, the insensible but none the less real future, the yonder, draws us onward, away from the here which is the only place we can truly be. Consequently we are never quite fully here. Nor do we expect to be. Every venture into the beyond opened up by indications opens up another beyond. Dune after dune of possibility recedes to an horizon that in turn recedes as we advance, until we imagine the ultimate beyond which lies neither over here nor over there, a beyond beyond which there is no beyondering.

At this point, we leave the index finger and the world it has helped to open up by a mode of reaching in which the hand transcends itself: a grasp that goes beyond prehension to apprehension and comprehension. The joint attention required for, and elaborated in, pointing illustrates how we humans get together in a way that no other creature does. Our powers, our range and the complexity of our lives are magnified as, by virtue of a profound sharing that reaches to the heart of our being, we translate our amassed numbers into accumulated knowledge and expertise, so that the potency of one human partakes of the power of all humanity. How small the index finger is and how great its effect. When it joins with other index fingers, and with systems of signs inscribed not in flesh but in air and ink, for which it prepares the way, it allows us to live, in the interval between our biological birth and our biological death, in a light other than that dispensed by the sun, in a space unknown to nature.

Notes

Foreword

1. Raymond Tallis, *The Hand. A Philosophical Inquiry into Human Being* (Edinburgh, Edinburgh University Press, 2003).
2. Comprising Raymond Tallis, *The Hand*, ibid.; *I Am. A Philosophical Inquiry into First-Person* (Edinburgh, Edinburgh University Press, 2004); and *The Knowing Animal. A Philosophical Inquiry into Knowledge and Truth* (Edinburgh, Edinburgh University Press, 2005).
3. L. E. Jones-Engel and K. A. Bard, 'Precision grip in young chimpanzees', *American Journal of Primatology*, 1996, 35, pp. 1–5.
4. Raymond Tallis, *The Kingdom of Infinite Space. A Fantastical Journey Around Your Head* (London, Atlantic, 2008).
5. Just after I had completed the second draft of *Michelangelo's Finger*, I came across Sotaro Kita, ed., *Pointing. Where Language, Culture, and Cognition Meet* (Hove, Psychology Press, 2008; first published 2003). I was at first dismayed at the thought that I had been scooped as a result of allowing my ideas on the subject to gestate for over thirty years. On reading this excellent book I was relieved to discover that its emphasis was different from my own approach and preoccupations, which are best described as 'philosophical anthropology'. As my worry evaporated,

I realized I had had a bit of luck. Kita's book has proved an invaluable resource in this final writing, in those areas where I make or imply empirical assertions (such as, for example, the universality of pointing in humans and its absence in animals, and the way pointing 'works'), as will be evident in the frequency with which I cite contributors to that volume.

Chapter 1: How to Point: A Primer for Martians

1. *The Confessions of Saint Augustine*, tr. E. B. Pusey (Glasgow, Blackie and Son [1904]), p. 12.
2. Cratylus was a follower of Heraclitus, who famously argued that, because everything is in a state of flux, you could not step into the same river twice. Cratylus carried this doctrine of flux to its logical conclusion and concluded that you could not step into the same river even once, since during the process of stepping into the river, the latter would have changed. According to Aristotle (*Metaphysics*, 1010a), Cratylus also argued that since 'regarding that which everywhere in every respect is changing nothing could be truly affirmed' and the right course is just to stay silent and wag one's finger.
3. David Wilkins, 'Why Pointing With the Index Finger is Not a Universal (in Sociocultural and Semiotic Terms)', in Kita, op. cit., pp. 171-215.
4. For a brilliant account of reaching and the control of the hand, see Allan M. Wing, Patrick Haggard and J. Randall Flanagan, eds, *Hand and Brain. The Neurophysiology and Neuropsychology of Hand Movements* (London, Academic Press, 1996). My *The Hand*, op. cit., examines this from a more philosophical viewpoint.
5. Even without the assistance of words, putting one's self in another's place does not always occur explicitly. When you point something out to me, I do not consciously adopt the viewpoint of your body. I simply look 'over there' to where you are pointing. But I can take this short cut only after I have already acquired the skill that enables me, as it were, to *triangulate* between you, the object and me. This is possible only on the basis of the initial sense of where you are pointing from; of dis-placing myself and putting you in my place.
6. This is relevant to our consideration of animal pointing.

Chapter 2: What it Takes to be a Pointer

1. What follows is expounded at much greater length in my *I Am*, op. cit., and *The Knowing Animal*, op. cit.
2. The best account of Merleau-Ponty's philosophy of the embodied subject is in Eric Matthews, *The Philosophy of Merleau-Ponty* (Montreal and Kingston, McGill/Queen's University Press, 2002).
3. G. G. Gallup, 'Self-recognition in primates. A comparative approach to the bi-directional properties of consciousness', *American Psychologist*, 1977, 32, pp. 329–38.
4. For a discussion of this in relation to our own heads, see Raymond Tallis, *The Kingdom of Infinite Space*, op. cit.
5. See, for example, Lecture III: 'Non-conceptual Content', in John McDowell, *Mind and World* (Boston, MA, Harvard University Press, 1994).
6. I have argued in *The Hand*, op. cit., especially in Chapter 9, 'The Tool of Tools', that the further generalization of the self, and the collectivization of consciousness, beyond what is made available by pointing, is crucially driven by the development of tools, themselves inspired by the 'toolness' of the hand and the instrumentalization of the body. Tools are extracorporeal manifestations of generalized human need, purposes, agency and so on. Primitive tools are shared in common. They also serve a multiplicity of social (often religious) functions. This combination of characteristics makes them key elements in the building of a common world that is not merely a nexus of public objects but a nexus of public meanings that are the necessary precursors of language.
7. 'Particular actuality' may be thought of as a pleonasm, inasmuch as only possibilities can be general while the actual must be particular. In fact, as we have seen, even the objects of indexical awareness are loci of possibility, inasmuch as they are not exhausted by their appearances. Here we are close to the intuitions that prompted J. S. Mill to describe 'matter' and 'material objects' as 'the permanent possibility of sensation'.
8. Such as that the world was a single, homogeneous, unchanging block and that the future and the past were illusions. See Raymond Tallis, *The Enduring Significance of Parmenides. Unthinkable Thought* (London, Continuum, 2007).
9. *See The Knowing Animal*, op. cit., Sections 4.1–4.3.

10. In infant development, there appear to be two stages in realizing that others see things that we do not see. According to Flavell and colleagues, children appear by two to three years of age to appreciate that visual perception connects people to external, public objects or events. It is not until they are four or five that they fully understand that seeing is associated with a particular internal viewpoint – an advantage or disadvantage point – within the world, though this is implicit in pointing. This second stage marks the stage at which children start to understand that seeing causes internal knowledge states in people – both one's self and others. J. H. Flavell, B. A. Everett, K. Croft, E. R. Flavell, 'Young children's knowledge about visual perception: further evidence for the level 1–level 2 distinction', *Developmental Psychology*, 1981, 17, pp. 99–103.

Chapter 3: Do Animals Get the Point?

1. Nicholas Humphrey, Foreword to Daniel J. Povinelli, *Folk Physics for Apes. The Chimpanzee's Theory of How the World Works* (Oxford, Oxford University Press, reprinted with corrections, 2003), p. v.
2. For brief accounts of Darwinosis and an even briefer critique, see Raymond Tallis, 'Carpal Knowledge: A Natural Philosophy of the Caress', *Philosophy Now*, September/October 2001, pp. 24–7; and 'Escape from Eden', *New Humanist*, November 2003, pp. 22–4. Anyone with an appetite for a longer account may wish to consult *The Hand*, op. cit., and *The Knowing Animal*, op. cit.
3. To be fair to pet-lovers, and to today's Darwinotics, we should acknowledge that Darwin himself also believed that 'there was no fundamental difference between man and higher mammals in their mental faculties' (*The Descent of Man*, cited in Povinelli, op. cit., pp. 10–11). He asserted that the difference between the minds of humans and other animals was 'certainly one of degree and not kind'. Just in case there is any misunderstanding, he spells this out in detail: 'It has, I think, now been shewn that man and the higher animals, especially the Primates...[a]ll have the same senses, intuitions, and sensations, – similar passions, affections and emotions, even the more complex ones, such as jealousy, suspicion, emulation, gratitude and magnanimity; they practise deceit and they are revengeful; they are sometimes susceptible to ridicule,

and even have a sense of humour; they feel wonder and curiosity; they possess the same faculties of imitation, attention, deliberation, choice, memory, imagination, the association of ideas, and reason, though in very different degrees' (ibid.).

4. Povinelli, op. cit., pp. 50–54.

5. As we shall see in the next chapter, it takes no time for infants to graduate from imperative to informative – declarative, genuinely referential – pointing.

6. These gestures are most likely developed by operant conditioning reinforced by rewards. The evidence overwhelmingly suggests that chimpanzees neither understand them nor intend them in the way that humans do (Povinelli, op. cit., pp. 53–4).

7. Chimps are not, of course, entirely void of self-consciousness and of being their bodies. As already mentioned, unlike gorillas and gibbons they recognize themselves in mirrors. There is, however, considerable argument as to whether the self-concept this implies is a representation of a *psychological* self or what Povinelli calls a *kinaesthetic* self-image, based upon the fact that when the chimp's body moves, its image moves. Povinelli argues that when a chimp looks in a mirror, unlike a child, he concludes 'That's the same as me!' rather than 'That's me!' In fact, it would be difficult to conclude the former without having a sense of the former. My own inclination is to characterize the difference as follows: a chimp's self-consciousness is fleeting, intermittent and non-cumulative, while human self-consciousness is an infusing sense that grows stronger through infancy, remains constantly present and accretes ever-more layers, all illuminated by, or underwritten by, the existential intuition. For more on this, see Tallis, *I Am*, op. cit.

8. J. Goodall, *The Chimpanzees of Gombe. Patterns of Behaviour* (Cambridge, MA, Belknap, Harvard University Press, 1986).

9. It is tempting to be more precise and say that pointing is a mode of singular, definite, reference. This would be to go too far because one can have 'singular' and 'definite' reference only if it is also possible to have general, indefinite reference. For reasons which we shall go into in Chapter 5, this is not possible.

10. John Lyons, *Semantics* (Cambridge, Cambridge University Press, 1981), vol. 1, p. 174.

11. Herbert Clark, 'Pointing and placing' in Kita, op. cit., pp. 243–68.

12. Povinelli, op. cit., p. 9.

Chapter 4: People Who Don't Point

1. We have to be careful here not to over-interpret what is going on, as we have already discussed. This is an issue to which we shall return in the next chapter.

2. E. Bates, L. Camaioni, V. Volterra, 'The acquisition of performatives prior to speech', *Merrill-Palmer Quarterly*, 1975, 21, pp. 205–224.

3. Victoria Southgate, Catherine van Maanen and Gergely Csibra, 'Infant Pointing: Communication to Cooperate or Communication to Learn?', *Child Development*, 2007, 78, pp. 735–740.

4. M. Tomasello, M. Carpenter, U. Liszkowski, 'A New Look at Infant Pointing', *Child Development*, 78, pp. 705–722.

5. Southgate, op. cit.

6. A. Lock, A. Young, V. Service and P. Chandler, 'Some observations on the origins of the pointing gesture', in V. Volterra and C. J. Erting, eds., *From Gesture to Language in Hearing and Deaf Children* (Berlin, Springer Verlag, 1990), pp. 42–55.

7. Charlotte Moore, *George and Sam* (London, Viking, 2004), p. 34. I am more indebted to this wonderful, brave book than is reflected in the specific citations.

8. Ibid., p. 37.

9. Ibid., p. 3.

10. Uta Frith, Autism: *Explaining the Enigma* (Oxford, Blackwell Science Publications, 1989).

11. No wonder autists are startled so often, and are so much more affected, by certain sensory experiences – for example, by thunder: they cannot tidy them away into conceptual drawers. The sense of lack of control, of being threatened by an engulfing reality, prompts much behaviour aimed at controlling the world, in reality or symbolically, by routines that are departed from on pain of hysteria and frightening tantrums.

12. This is accessibly summarized in Povinelli, op. cit., pp. 308–310.

13. Moore, op. cit., p. 29.

14. Tallis, *I Am*.

15. Moore, op. cit., p. 28–9.

16. Povinelli, op. cit., p. 21.

17. Ibid., p. 49.

18. We could elaborate this further. Autists, who seem to be more pinned to the here-and-now – their attention seems to be fastened to the perceptual field like a burr on Velcro – could not assume the distance that would enable them to point out their noticings in order to bring them to the notice of another who inhabits a different viewpoint. Autists stick so closely to the point, to what is the case, that they cannot be uncoupled from the here and now in order to point out what is the case. Imprisonment in the here-and-now blocks expansion into the world of the invisible, of the hidden minds of others and the hidden properties and places of their world.

Chapter 5: Pinning Language to the World

1. Cited in Nobuo Masataka, 'From Index-Finger Extension to Index-Finger Pointing: Ontogenesis of Pointing in Preverbal Infants', in Kita, op. cit., p. 82.

2. It is interesting to note in this context that the pincer grip – between forefinger and and thumb – is systematically selected in infants only a month before pointing occurs. Prior to that there is 'tipping', exploration with the tip of the finger. It is almost as if pointing, unlike tipping, is a virtual grasping, a kind of prehension at a distance, a pincer grip that withholds itself. To point at something rather than grabbing it brings touch closer to the fastidious distance of vision.

3. This brief survey is indebted to many sources but most of all to the excellent editorial introduction by G. H. R. Parkinson to *The Theory of Meaning* (Oxford, Oxford University Press, 1968).

4. See Tallis, 'Some truths about truth', in *The Knowing Animal*.

5. Daniel Everett, *Don't Sleep, There are Snakes. Life and Language in the Amazonian Jungle* (London, Profile, 2008).

6. W. E. Johnson's *Logic* was published in three and a half parts between 1921 and 1932. For an account of this remarkable publication, see John Passmore, *A Hundred Years of Philosophy* (London: Penguin, 1968), pp. 343–6. His lack of publishing would have ensured he did not survive in contemporary academe. The Research Assessment Exercise would

probably have relegated him to teaching or domestic duties, or driven him to early retirement.

7. Robert Audi, ed., *Cambridge Dictionary of Philosophy* (Cambridge, Cambridge University Press, 1995), p. 186.

8. These alternative methods are no less vulnerable to misunderstanding and no less dependent on assumptions than is the canonical method achieved by pointing. Like pointing (as we shall discuss presently), actions such as pinching to define pinching may be misinterpreted as an ostensive definition of say, 'skin' or 'pain'. Interestingly, ostensive definition may also contribute to defining itself, as when we point to an object and give it its names as an illustration of how ostentive definition works.

9. John Searle, *Speech Acts. An Essay in the Philosophy of Language* (Cambridge, Cambridge University Press, 1969), p. 162.

10. Ibid., p. 174. Searle is here echoing Plato's argument with Cratylus, one of the deep interlocutors of this book. (See n. 2, p. 144.) If following Cratylus you cannot step into a river even once, nothing can be truly said of any given object – we cannot predicate anything of a (stable, underlying) subject, and all we can do is to refer to objects by pointing at them. Plato argued that Cratylus could not do even that because reference requires predication.

11. That is why, as mentioned in Chapter 3, we should not even suggest that pointing is closer to one form of reference – singular, definite, reference – than another, say, reference to a class of objects. One can have 'singular' and 'definite' reference only if it is also possible to have general, indefinite reference.

12. Anthony Quinton, 'The *A Priori* and the Analytic', *Proceedings of the Aristotelian Society*, 1963–4, 64, pp. 31–54. Collected in P. F. Strawson, ed., *Philosophical Logic* (Oxford, Oxford University Press, 1967), pp. 107–128. The page references in the text are to the reprinting in Strawson.

13. I have elsewhere (and more accurately) called these aspects 'sense ions'. See Raymond Tallis, *Not Saussure. A Critique of Post-Saussurean Literary Theory* (London, Macmillan, second edition, 1995), especially Chapter 10.

14. It is not possible to point to '-ness'. The notion that recurrent pointings

to a succession of objects sharing the same property will generate '-ness' is not terribly persuasive, either.

15. Ludwig Wittgenstein, *Philosophical Investigations*, tr. G. E. M. Anscombe (Oxford, Blackwell, 1958), p. 2e.

16. *Vide* his remark in *On Certainty,* ed. G. E. M. Anscombe and G. H. von Wright, tr. Denis Paul and G. E. M. Anscombe (Oxford, Basil Blackwell, 1969), Para 141, p. 21e: 'When we first begin to *believe* anything, what we believe is not a single proposition, it is a whole system of propositions. (Light dawns gradually over the whole.).'

    Hegel had already pointed out that one could not establish a referent by pointing alone. One cannot point to a particular without first specifying what it is to which one is pointing. See Robert C. Solomon, 'Hegel's *Phenomenology of the Spirit*', in Robert C. Solomon and Kathleen M. Higgins, eds., *The Age of German Idealism, Routledge History of Philosophy*, vol. 6 (London, Routledge, 1993).

17. Ray Jackendoff, *Foundations of Language. Brain, Meaning, Grammar, Evolution* (Oxford, Oxford University Press, 2002), p. 239. None of this is found in primate calls, although those who are desperate to prove that we are just clever chimps stubbornly overlook this.

18. Another reason for being cautious about attributing too great a part to ostension in mediating between our speaking selves and the material world is that ostension has several possible roles. We may use ostension: a) to baptise something – 'I name this colour red'; b) to identify something – 'What is that colour called?'; c) to capture the meaning of a word – 'What is the meaning of red?' Because it can seem to do all of these things, depending where you are starting from, it is evident that in truth it does none of them by itself. It is, incidentally, deeply mysterious that there should be a separation of these functions – of baptising, identifying (classifying) and answering a query as to the name of the object – of the act of naming and the use of the name and the explanation of the use of the name.

    The most extraordinary example of supposed ostensive definition is to be found in a lecture given by the philosopher G. E. Moore when he claimed that he had proved the existence of the outside world by demonstrating two objects existing in it: his own hands. His demonstration was in fact more of a 'monstration': he raised up his hands for his audience to see. This was an ostensive definition of existence, though it hovered uneasily between a definition of the word 'existence' (as illustrated in two existent beings) and a proof that there were existents. (For the full grisly story see Raymond Tallis, 'George Moore's Hand: Scepticism About Philosophy', *Philosophy Now*, 2008, 69, pp. 48–9.)

19. Nonsense that I have examined in great detail in *Not Saussure*, op. cit., and *In Defence of Realism* (Lincoln, NE, University of Nebraska Press, second edition, 1998).

20. Genius does not guarantee immunity from this delusion. It was revived in the 1970s by the great American logician and philosopher Saul Kripke, who incidentally published his first serious paper in the philosophical literature when he was in his teens.

Kripke introduced the concept of 'a rigid designator' as a way of supporting a particular position he had on the nature of identity and the relationship between necessity and empirical truths. This is no place to rehearse his theory about the essential necessity of identity. Interested readers may consult *Naming and Necessity* (Oxford, Basil Blackwell, 1980). It is, however, much to our purpose to focus on one aspect of it.

Kripke argues that many truths that have been discovered only as a result of empirical inquiry are necessary truths, in contradiction to the venerable belief that all necessary truths are a priori and independent of sense experiences. Such truths include identity relations, for example that the Morning Star is the Evening Star. The 'two' stars are in fact different manifestations of a single star, namely Venus: the Evening Star is Venus in the evening and the Morning Star is Venus in the morning. This was, however, discovered only a couple of hundred years ago. The identity was established as a result of an empirical observation, and yet it was necessarily true: there is no possible world in which Venus could be non-identical with Venus. (There have been many critiques of this argument. My own is 'Saul Kripke Tries to Brighten Things Up', in *I Am*, op. cit.)

The seeming paradox of a necessary truth being established by an empirical observation arose, Kripke argued, because the star Venus was a referent that could be accessed in different ways, through different *senses* – 'Morning Star', 'Evening Star' and 'Venus'. He then put forward the notion of a 'rigid designator', through which an object could be accessed, which was immune from the confusing variations in sense that had enabled the error of thinking the Morning Star and the Evening Star were two different stars to persist even after science had shown otherwise. During its history, an object might acquire many accretions of sense by means of which it might be accessed. For example, Aristotle might be known as the teacher of Alexander, yet it is not impossible that he had lied and perhaps had not even met Alexander. Even so, this would not make him a different person. So we need a mode of designating, or pinning down (the phrase is inescapable) an object so directly that it is not subject to the vagaries of sense (and of what people know about the object). Whatever is true of the object accessed this way would be true in all possible worlds and hence necessarily true. This is the object captured in its singularity through a rigid designator. The rigid designator owes its status to an act of *original baptism*, whereby the object is named. And this act of baptism is, once more, an ostensive definition: the object is pointed to and named. Ostensive definition seems to access the object purely and capture its naked identity denuded of accretions of sense.

We can now see how attractive that idea is and, at the same time, how wrong it is. There cannot be an act of ostensive baptism that then sends an object into the world of discourse armour-clad in its material singularity against the encroachment of secondary senses through which it has to be accessed. Once an object is captured for language, it has to play by the rules. It becomes a part of speech and its singularity can be accessed only by playing the linguistic game. It cannot be prevented from participating in the confusions that result from being absorbed into the medium of language.

21. Butterworth, op. cit., in Kita, op. cit., p. 27.
22. Meredith L. Rowe and Susan Goldin-Meadow, 'Differences in Early Gesture Explain SES Disparities in Child Vocabulary Size at School Entry', *Science*, 2009, 373, pp. 951–3.
23. Butterworth, op. cit., in Kita, p. 29. Some of the 'cognitive metalling' is described in Chapter 2. A key element is beautifully described by Povinelli (in Kita, p. 66) when he tries to capture the background against which pointing does its business: 'The ability to represent others as possessing internal, attentional states may have allowed existing gestural

forms to be woven into a much larger, more efficient and productive system of communication – a system in which understanding the psychological meanings behind the gestures is just as important as the gestures themselves.' We are ripe to move on from pointing to speech because of what lies behind pointing. That is why gestural theories of the origin of language capture only half the truth. Yes, pointing is a proto-referential gesture, but behind pointing and true (verbal) reference is something that makes both the communicative gesture and speech possible. The something, that we examined in Chapter 2, is our relationship to our own body, which yields a sustained sense of self so that we are not mere organisms but embodied subjects.

Chapter 6: Pointing and Power

1. H. W. Fowler, *A Dictionary of Modern English Usage,* ed. and rev. Sir Ernest Gowers (Oxford, Clarendon Press, 1968), 'One'.
2. For a discussion of this see Tallis, *In Defence of Realism*, pp. 64–6.
3. One of the most amusing and penetrating explorations of the relationship between the forefinger and the accusation of crime is to be found in Charles Dickens' *Bleak House*, where he discusses the suspicions of the detective Mr Bucket:

> Mr Bucket and his fat forefinger are much in consultation together under existing circumstances. When Mr Bucket has a matter of pressing interest under his consideration, the fat forefinger seems to rise, to the dignity of a familiar demon. He puts it to his ears, and it whispers information; he puts it to his lips, and it enjoins him to secrecy; he rubs it over his nose, and it sharpens his scent; he shakes it before a guilty man, and it charms him to destruction. The Augurs of the Detective Temple invariably predict that when Mr Bucket and that finger are much in conference, a terrible avenger will be heard of before long.
>
> …[Mr Bucket] is in the friendliest condition towards his species, and will drink with most of them. He is free with his money, affable in his manners, innocent in his conversation –

but, through the placid stream of his life, there glides an
undercurrent of forefinger.

Charles Dickens, *Bleak House* (Harmondsworth, Penguin, 1985), p. 768.

4. T. S. Eliot, 'The Love Song of J Alfred Prufrock', in *Collected Poems,
   1909–62* (London, Faber and Faber, 1963), p. 15.

5. See J.-P. Sartre, 'The Look', in *Being and Nothingness. An Essay on
   Phenomenological Ontology*, tr. Hazel Barnes (London, Methuen, 1957),
   pp. 252–302.

6. G. F. W. Hegel, *The Phenomenology of Spirit*, tr. A. V. Miller (Oxford,
   Oxford University Press, 1977), pp. 175, 178.

7. The argument behind this argument can be made immensely
   complicated: see Tallis, I Am, especially Chapter 2.

8. See Raymond Tallis, 'Identity and the Mind', *Identity. Darwin College
   Lectures 2007* (Cambridge, Cambridge University Press, in press).

9. Anton Chekhov, 'A Boring Story' in *The Complete Short Novels*, trs.
   Richard Pevear and Larissa Volokhonsky (London, Vintage, 2005).

10. Peter Godwin, *When the Crocodile Ate the Sun* (London, Picador, 2007), p.
    165.

11. Tallis, *In Defence of Realism*, p. 122.

Chapter 7: Assisted Pointing and Pointing by Proxy

1. In our Martian's primer on pointing we referred to the instrument used
   to point with – the finger, the rod, etc. – as the 'pointer'. The person
   doing the pointing we called 'the producer'. This is a little awkward, and
   we might feel it more just to describe the *person* as the pointer: isolated
   fingers do not point; people use fingers to point. That is perfectly correct,
   but humans have an inveterate tendency to attribute to those things we
   use to convey meanings, or indeed to bring about other effects, the
   intentionality or intentions of those who use them. This tendency is
   carried beyond the body when we talk about non-carnal pointers, those
   entities that assist pointing. This has had a huge and misleading effect on
   much thinking about minds, machines and many other things, and it has
   enabled otherwise implausible materialist theories of mind to command
   acceptance. Materialism borrows meaning from humans and inserts it
   into configurations of matter, which are then treated as signs, symbols

and other bits of consciousness. This tendency also played a role in the displacement of the philosophy of consciousness by a philosophy focussing on concepts. And it is central to the computational theory of mind. The interested reader might want to consult my 'Pocket Lexicon of Neuromythology' called *Why the Mind is Not a Computer*, (Exeter, Imprint Academic, 2005).

2. Cynics have suggested that the primary purpose of visual aids is to distract the audience from the intellectual nullity of the lecture. As Professor Henry Miller, one-time Vice-Chancellor of the University of Newcastle-upon-Tyne, remarked to me after I had given a talk to which I had paid considerable attention to the A-V side of things, there are two sorts of talks: those with visual aids; and those with original thought.

3. The casual, effortless metaphor by which occulting lights are called winkers should not be underappreciated. Bodily winking does not, of course, extinguish a source of light, but rather denies a recipient the light it would otherwise harvest. This is an interesting inversion, as is the fact that winking artefacts can wink on as well as wink off.

4. There seems to be no limit to the versatility of iconic signs. The use of a straight arrow to indicate a direction in space, in time, or of a succession of processes is in daily use on keyboards everywhere – the 'Enter' key has an arrow indicating down and then to the left.

5. See, for example, K. R. Popper, 'On the Theory of the Objective Mind', in *Objective Knowledge. An Evolutionary Approach* (Oxford, Oxford University Press, 1972).

6. W. H. Auden, *Collected Poems* (London, Faber and Faber, 2007), 'In Time of War', Section VIII.

7. William Shakespeare, *Hamlet*, II:2, 1660.

8. There is a difference of opinion between the two dictionaries as to how the word should be spelt. The American *Webster's* says 'fingerpost' while the British *OED* gives 'finger-post'. British usage in general is more comfortable with hyphens; America-in-a-hurry bothers less with them. We could regard the hyphen as a wonderfully compressed image of the Atlantic Ocean or a bridge between the diverging histories of two nations who have rather a lot of past.

9. This is a concrete use of written language that really does work. While concrete poets have often played with the spatial disposition of written

words in the hope of realizing their subjects both referentially and iconically, this rarely does anything for the reader. The most dramatic examples – for example Apollinaire's poem about the rain, in which the words pour down the page on their sides – works as an idea, as a visual conceit, but not as a poem. By the time one has turned the page on its side to decipher the poem, the idea has modulated into an impediment to understanding.

10. Friedrich Nietzsche, 'On Truth and Lie in an Extra-Moral Sense', tr. Walter Kaufmann, in *The Portable Nietzsche* (New York, Viking Press, 1954), p. 42.

11. Thomas Stanley, *Historical Memorials of Canterbury Cathedral* (1857), i:31; Thomas Beddoes, *Observations on the Nature of Demonstrative Evidence* (1797), p. 138. It is interesting in this context to remind ourselves how so-called 'deictic shifters' – pronouns such as 'I' and 'me' which change their reference depending on who is uttering them – are indicated in sign-language by pointing.

Chapter 8: The Transcendent Animal: Pointing and the Beyond

1. Martin Heidegger, *What is Called Thinking?*,tr. F. D. Wieck and J. G. Gray (New York, Harper & Row, 1968), pp. 148–9.

2. W. D. Ross, ed., *Aristotelis De Anima* (Oxford, Clarendon Press, 1956), 424a, pp. 17–19.

3. John Maynard Keynes, 'Economic possibilities for our grandchildren', in *Essays in Persuasion. The Collected Writings of John Maynard Keynes* (London, Macmillan, Cambridge University Press for the Royal Economic Society, 1972), vol. 9, pp. 323–4.

4. This is from Barry Allen's brilliant and neglected *Knowledge and Civilization* (Boulder, CO, West View, 2004), p. 130.

5. Ibid., p. 130. This is Allen's summary of an argument from Michel Foucault.

6. Jean-Yves Lacoste, 'Perception, Transcendence and the Experience of God', in Peter M. Candler and Conor Cunningham, eds., *Transcendence and Phenomenology* (London, SCM Press, 2007), p. 18.

7. W. B. Yeats, 'The Circus Animals' Desertion', in *Yeats's Poems*, A. Norman Jeffares, ed. (Basingstoke, Macmillan, 1996), p. 472.

Index

accusation, and pointing 93
actuality, particular 32–4
agency *see* self-consciousness
air-jabbing 90–1
Allen, Barry 137–8
Alston, William 68
Althusser, Louis 92
animals:
    and bodily postures 25
    and classifying behaviour 125
    and cognition 38, 43, 147 n.3
    and communication 49
    failure to teach young 44, 77, 118
    inability to point xix, 16, 37–50,
    118, 145 n.6
    and the invisible 47, 59
    and self-consciousness 18–19, 35,
    39, 49
    and sentience 26–7, 39, 48
Aristotle xv, 22, 124, 126, 127, 140,
145 n.2

arm, role in pointing 5–6, 8, 104
Asperger's syndrome 39
assertion:
    and animals 47–9
    and pointing 47
    and truth 46, 48
attention:
    directing 4, 12–13, 15, 34, 36,
    45–6, 50
    joint 10, 52, 60, 114, 117, 119,
    131, 143
Auden, W. H. 110
Augustine of Hippo, St 1, 77–9, 84
Austin, J. L. 8, 68
autism:
    and awareness of other minds
    11–12, 55, 60
    and bodily awareness 58–9, 61
    and embedded-figure test 58
    and failure to point xix, 11–12,
    54–5

power grips xvi

predication, and reference 46, 47–8, 71

primates:

    and Darwin 147–8 n.3

    and failure to point 41–4, 47–8, 118

    and fractionated finger movements 6–7, 22

    and gaze-following 43, 59–60

    and other minds 43–4, 47

    and self-consciousness 18–19, 24, 50, 148 n.7

producer:

    and awareness of other minds 11–13, 18, 42, 118

    and body as communicative tool 10–11, 15–16, 18, 24–5

    and meant meaning 11, 12

    and mental attention 13

    and pointee 29, 105–6

    as pointer 133, 156 n.1

    and rules of pointing 7–11

proper names 71, 74, 75, 112

Quinton, Anthony (Lord Quinton) 72–4

Raphael, *The School of Athens* 140

reality:

    collective 55, 114, 118–19, 129–33

    hidden 119, 123, 129, 131, 139

    and language 62–9, 80–3

reference:

    indefinite 148 n.9, 151 n.11

    and language 45–6, 85, 131

    and meaning 75–6

    by pointing 4, 39, 41–3, 44–6, 47–8, 83, 118

    and predication 46, 47–8, 71

    proto-reference 44-6, 47-8, 71, 75

    self-reference 98–101

reflexiveness 27–8

rudeness of pointing xx, 93–8

Sartre, Jean-Paul 94

Searle, John 71

self:

    and autism xv, 12, 56, 58, 60

    and identity 96–8

self-consciousness:

    and animals 18–19, 35, 39, 49, 96

    and other consciousnesses 95–8, 101

    and pointing 104–5

    and primates 18–19, 24, 148 n.7

    as uniquely human xviii, 18–20, 95–6

    and viewpoints 20, 23, 26, 31, 114, 118, 131

    see also consciousness

self-reference, and pointing 98–101

sensation, and perception 120–2

sense experience see sentience

sentience, and animals 26–7, 39, 48

Shakespeare, William, *Hamlet* 110

sign:

    index finger as 45

    and meta-language 99–101

signalling, manual 107

signposts, as proxy pointers 14, 109–16

silence, and finger on lips 90

Southgate, Victoria, van Maanen, Catherine and Csibra, Gergely 54
space, common 28, 118, 130–2
speech, infant, and pointing 83–6, 131
speech act theory 68, 82
Stanley, Thomas 115, 158 n.11
structuralism 80–1
subject, embodied xv, 18, 20, 21, 23–4, 26, 35, 47, 118
    and animals 27, 39–40
    and autism 58–9
summons, pointing as 91

thumb:
    opposable xv, xviii, 21
    as pointer 103
Tomasello, M., Carpenter, M. and Liszkowski, U. 53–4
tools:
    and collectivization of consciousness 146 n.6
    hands as xv, xviii, 16, 21–2, 24–5, 41, 103–4, 119
touch 22, 23, 120–7
    and gaze 94, 123
    virtual 91, 106, 129
transcendence:
    and consciousness xx, xxii, 9–10, 119–43
    of the divine xx, 139–43
    of perception 121–4, 126, 139

as possibility 130, 132, 135
    as public 130
truth and falsehood 32–4, 36, 46
    and meaning 67
    and possibility 33, 48

viewpoints:
    and autism 11–12, 55
    and child development 34–5, 52
    and primates 44, 47
    and self-consciousness 20, 23, 26, 31, 114, 118, 131
vision 122–32
    and classification 125–6
    and generality 129
    and the invisible 36, 123–4, 129, 139
    and pointing 21, 24, 47
visual aids, and pointers 106

Wilkins, David 2
Wittgenstein, Ludwig von 39, 62–3, 67, 74, 77–9, 81–2
words, as tokens 112–13
writing, finger-writing 98–100
Wundt, Wilhelm 64

Yeats, W. B. 142

Index compiled by Meg Davies
(Fellow of the Society of Indexers)